ELECTRON THEORY

ELECTRON THEORY

Prakash Trivedi

Oxford Book Company
Jaipur ▪ New Delhi

Published By

Oxford Book Company

Regd. Off. "Bony Residency", Gate No. 2
Opp. Tilak Public School, Vishweswariya Nagar,
Gopalpura Road, Jaipur - 302018, India
Ph: 0141 - 2761280, Telefax: 2761381
email - oxfordbookjpr@gmail.com
website - oxfordbookcompany.com

Delhi Office: 102, 1st Floor, Satyam House,
4327/3, Ansari Road, Darya Ganj,
New Delhi - 110002, India
Ph: 011-45652440

Published in India, 2025

ISBN : 9789380179704

Price : ₹ 2500/-

Typeset by Shivangi Computers, Jaipur

Printed and bound in India by Thomson Press Ltd.,
New Delhi

Preface

Electron theory states all matter is comprised of molecules, which in turn are comprised of atoms, which are again comprised of protons, neutrons and electrons. A molecule is the smallest part of matter which can exist by itself and contains one or more atoms.

To understand these processes requires an understanding of the basic principles. For the light to glow requires energy to find a path through the light switch, through the copper wire and this movement is called electron flow. It is also called current flow in electronics. This is the first important principle to understand.

An atom is also divisible - into protons and electrons. Both are electrical particles and neither is divisible. Electrons are the smallest and lightest and are said to be negatively charged. Protons on the other hand are about 1800 times the mass of electrons and are positively charged. Each are thought to have lines of forces (electric fields) surrounding them. In theory, negative lines of force will not join other negative lines of force. In fact they tend to repel each other. Similarly positive lines of force act in the same way.

Prakash Triveni

Contents

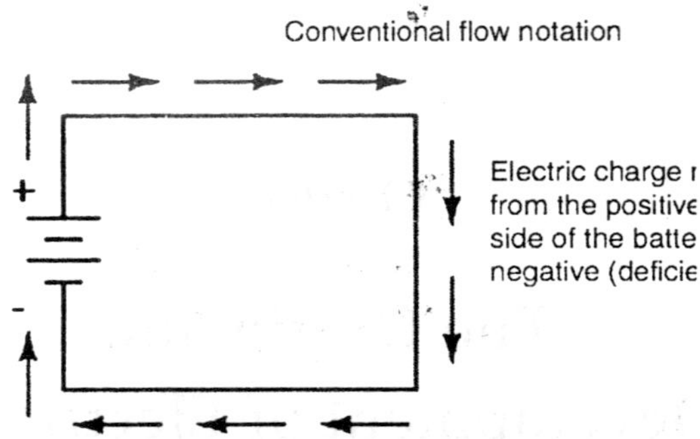

Others chose to designate charge flow according to the actual motion of electrons in a circuit. This is known as *electron flow* notation:

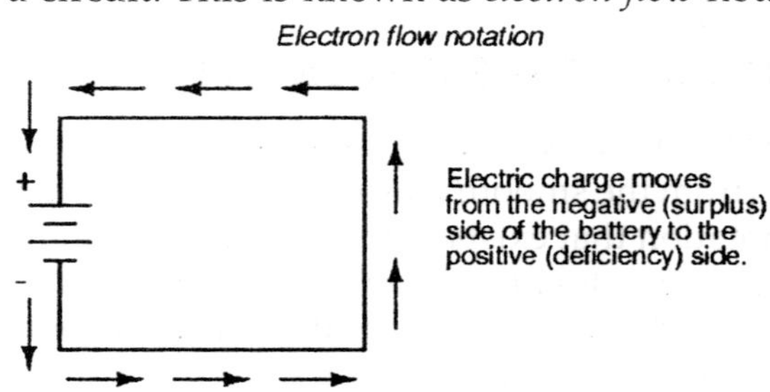

In conventional flow notation, we show the motion of charge according to the (technically incorrect) labels of + and -. This way the labels make sense, but the direction of charge flow is incorrect. In electron flow notation, we follow the actual motion of electrons in the circuit, but the + and - labels seem backward.

Does it matter, really, how we designate charge flow in a circuit? Not really, as long as we're consistent in the use of our symbols. You may follow an imagined direction of current (conventional flow) or the actual (electron flow) with equal success insofar as circuit analysis is concerned.

Many electrical devices tolerate real currents of either direction with no difference in operation.

Incandescent lamps (the type utilizing a thin metal filament that glows white-hot with sufficient current), for example, produce light with equal efficiency regardless of current direction. They even function well on alternating current (AC), where the direction changes rapidly over time. Conductors and switches operate irrespective of current direction, as well. The technical term for this irrelevance of charge flow is *nonpolarization.* We could say then, that incandescent lamps, switches, and wires are *nonpolarized* components. Conversely, any device that functions differently on currents of different direction would be called a *polarized* device.

There are many such polarized devices used in electric circuits. Most of them are made of so-called *semiconductor* substances, and as such aren't examined in detail until the third volume of this book series. Like switches, lamps, and batteries, each of these devices is represented in a schematic diagram by a unique symbol.

As one might guess, polarized device symbols typically contain an arrow within them, somewhere, to designate a preferred or exclusive direction of current. This is where the competing notations of conventional and electron flow really matter. Because engineers from long ago have settled on conventional flow as their "culture's" standard notation, and because engineers are the same people who invent electrical devices and the symbols representing them, the arrows used in these devices' symbols *all point in the direction of conventional flow, not electron flow*. That is to say, all of these devices' symbols have arrow marks that point *against* the actual flow of electrons through them. Perhaps the best example of a polarized device is the *diode*. A diode is a one-way "valve" for electric current, analogous to a *check valve* for those familiar with plumbing and hydraulic systems. Ideally, a diode provides unimpeded flow for current in one direction (little or no resistance), but prevents flow in the other direction (infinite resistance). Its schematic symbol looks like this:

Diode

Placed within a battery/lamp circuit, its operation is as such:

Diode operation

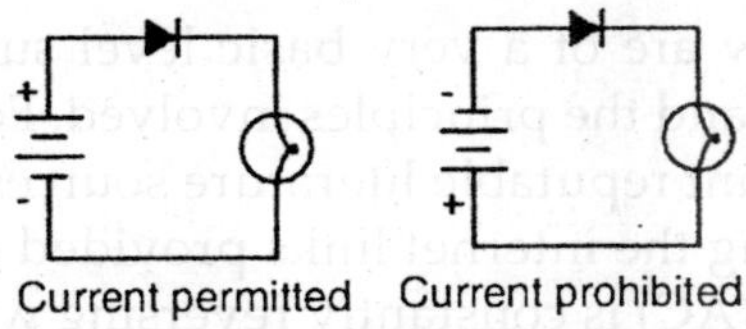

Current permitted Current prohibited

When the diode is facing in the proper direction to permit current, the lamp glows. Otherwise, the diode blocks all electron flow just like a break in the circuit, and the lamp will not glow.

If we label the circuit current using conventional flow notation, the arrow symbol of the diode makes perfect sense: the triangular arrowhead points in the direction of charge flow, from positive to negative:

Current shown using conventional flow notation

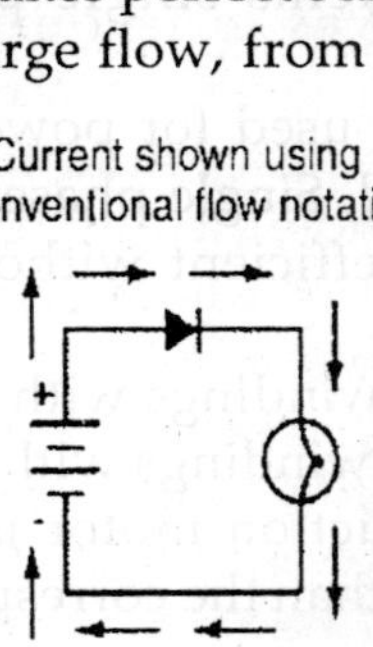

On the other hand, if we use electron flow notation to show the *true* direction of electron travel around the circuit, the diode's arrow seems backward:

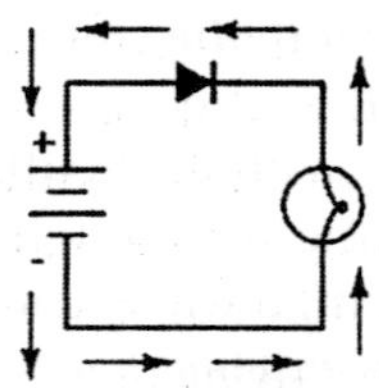

For this reason alone, many people choose to make conventional flow their notation of choice when drawing the direction of charge motion in a circuit. If for no other reason, the symbols associated with semiconductor components like diodes make more sense this way.

However, others choose to show the true direction of electron travel so as to avoid having to tell themselves, "just remember the electrons are *actually* moving the other way" whenever the true direction of electron motion becomes an issue.

ALTERNATING CURRENT THEORY

Introduction

The notes below are of a very basic level sufficient for mechanical engineers to understand the principles involved. For detailed information please refer to relevant reputable literature sources. Detailed information is also available using the internet links provided at the foot of this page. Alternating current (AC) is constantly reversing with a set frequency and in a generally sinusoidal pattern. Virtually 100% of industrial and domestic electric power supplies are alternating current.

Most high power electrical machines and appliances are based on AC supplies. In the UK the frequency of supply is 50 cycles/sec (Hertz).For lighting and heating systems single phase systems are perfectly satisfactory.

When AC supplies are used for powering induction motors single phase supplies are not ideal. Single phase induction motors are not self starting and are not very efficient without special auxiliary winding arrangements.

By using two separate windings with currents differing in phase by a quarter of a cycle or three windings with currents differing in phase by a third of a cycle, the induction motor is self starting and has better efficiency and power factor than the corresponding single phase machine.

Motors utilising two windings are called two-phase motors and motors with three windings are call three-phase motors.

These notes only discuss single phase and three phase systems. Notation -for factors used in rotating loop equations

A = Area of rotating loop in motor or generator (m^2) = w.l = w.2.r
B = Flux density (Tesla)
e = instantaneous e.m.f. (V).
E_m = maximum e.m.f generated (V)
f = frequency = (cycles/sec)
I = current (RMS) (amperes)
I_m= current (maximum) (amperes)
i = current (instantaneous) (amperes)
l = length of rotating loop (m)
r = radius of rotating loop (m)
w = width of rotating loop (m) = 2.r
n = speed of rotation of loop (revs/s)
u = velocity of sides of rotating loop (m/s)
V = Potential difference (RMS) (volts)
V_m= Potential difference (maximum) (volts)
v = voltage (instantaneous) (volts)
N = Number of windings
ω = Angular velocity of loop (rads/sec) = 2.π.f

Generation of Single Phase e.m.f.

At the most basic level a simple electrical generator as shown in principle below provides a sinusoidal voltage output.

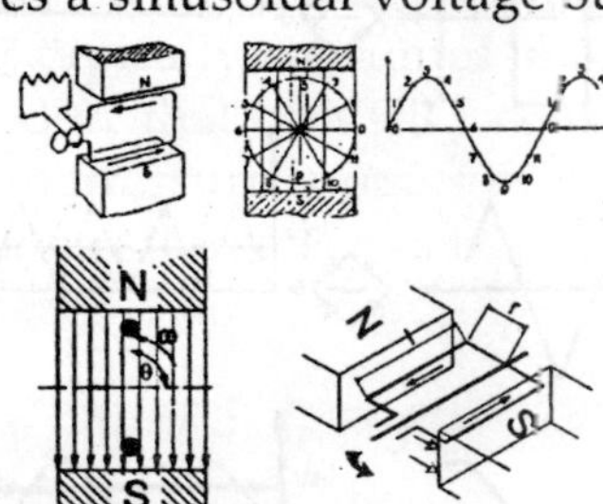

The e.m.f. generated in one side on the loop is B.l.u. sin θ. (Zερο voltage results when the side of the loop is moving parallel to the lines of magnetic flux and voltage = B.l.u when the velocity (u) is normal to the lines of flux.). The total emf generated by one side of the loop moving past the N pole + the other side of the loop moving past the S pole = 2.B.l.u sin θ.

The velocity u of the sides of the loop is ω.r and therefore the e.m.f. generated by the rotating loop at any angle =

$$e = 2.B.l.u.\sin\theta = 2.B.l.\omega.r.\sin\theta = B.A.\omega.\sin\theta = B.A.2.\pi.n.\sin\theta.$$

Note: A = area of covered by the loop = l.2.r

Alternatively if $E_m = B.A.\omega$ the instantaneous e.m.f. =

$$e = E_m \sin\theta = E_m \sin\omega.t$$

If the loop is replaced by a coil of N turns in series each having a area (2.r.l) = A then the maximum e.m.f. generated in the coil = E_m = B.A.2.π.n.N volts. The instantaneous e.m.f at any angle of rotation =

$$e = E_m \sin\theta = B.A.2.\pi.n.N \sin\theta$$

Average and R.M.S Values of sinusoidal

These notes relate primarily to cyclic variations of a variable. The principles also apply to variables which are constantly changing over time. There is a requirement to provide a single practical value to identify the magnitude of the variable (current, voltage etc) which varies in magnitude. Two methods of quantifying a representive value are available: "The Average Value" and "The Root Mean Square Value."

The Average Value

Intuitively the average value of the current or the voltage is the one which provides a useful practical measure of its magnitude. When an a.c. variable has been rectified the unidirecional current will deflect a d.c. meter. If the meter is has a suitable movement the meter will show the average value of the current. Three waveforms are shown below and the average value of their full wave rectified values are indicated.

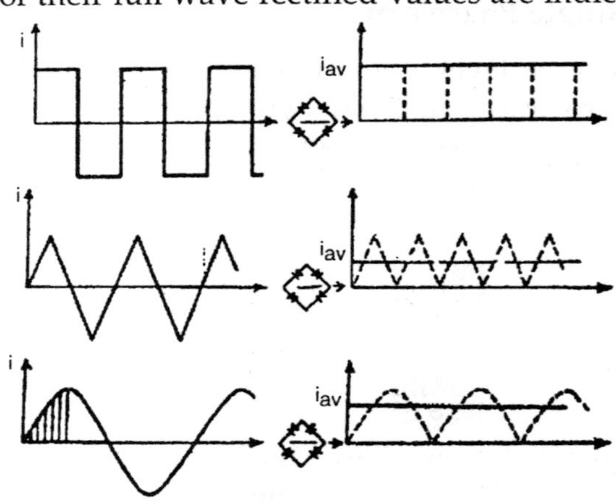

For the square wave pattern the average value of the full wave rectified form is $i_{av} = I_m$

For the triangular wage pattern the average value of the full wave rectified form is I_{av} = Im/2. For the sinusoidal wage pattern the average value of the full wave rectified form is $I_{av} = 0{,}637\ I_m$

Note: For all of these forms the average value for the half wave rectified version, when the negative part of the cycle is cut off, is half the value as shown above.

The Root Mean Square Value

In practical electrical engineering the average value is of little importance. It is most important to consider the work that the electricity can do either in heating or in mechanical work. This is generally related to the power equivalent of the current $W = I^2/R$ or V^2/R

Consider one cycle of the square waveform as shown below. (two cycles actually shown)

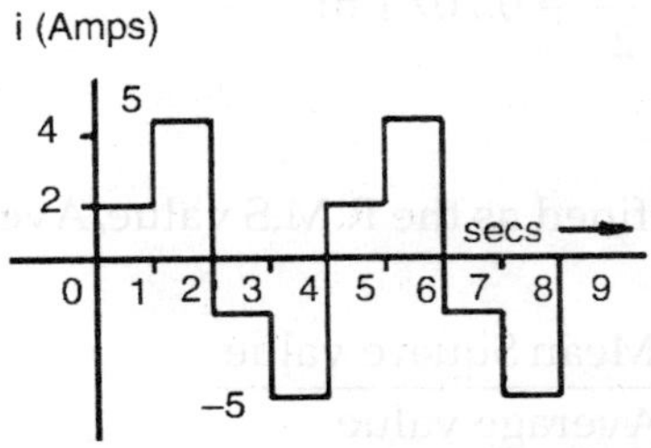

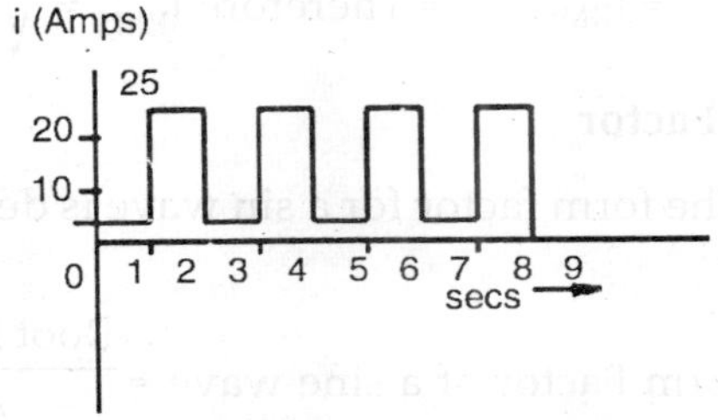

Assuming the current is delivering energy by heating a 4 ohm resistor. The total energy delivered over a complete cycle =

Energy = Σi^2 R.t = $[2^2.1 + 5^2.1 + -2^2.1 + -5^2.1].4$ =58.4 =232 Joules

The equivalent direct current to produce the same heating effect over a complete cycle = I.

This value of I called the Root Mean Square Value

I^2 R.t = =232 Joules

Therefore I = Sqrt (232/R.t) = Sqrt [232/(4.4)] = 3.807 Amperes

Note: The average full wave rectified value

= (2.1+5*1)*2/4 = 3.5 Amperes

For a sinusoidal current The RMS value for the current (I)

$I = 0{,}707.I_m$

(1) Average Value for a Ssinusoidal Variable

Consider a Sinusoidal current $i = I_m \sin\theta$ over a half Cycle...

$$\text{lav} = \frac{1}{\pi}\int_0^{\pi} i\,d\theta = \frac{1}{\pi}\int_0^{\pi} I_m \sin(\theta)\,d\theta\ \frac{1}{\pi} I_m\left[-\cos(\theta)\right]_0^{\pi}$$

$$= \frac{1}{\pi} I_m (1-(1-) = \frac{2}{\pi} I_m = 0.637\, I_m$$

2) Root Mean Square value for a sinusoidal variable

Consider a Sinusoidal current $i = I_m \sin\theta$.

$$l_{RMS}^2 = \frac{1}{2.\pi}\int_0^{2\pi} i^2 d\theta$$

$$= \frac{1}{2.\pi}\int_0^{2\pi} \sin^2(\theta)\, d\theta = \frac{1}{2.\pi}\frac{l_m^2}{2}\int_0^{2\pi}(1-\cos 2\theta).d\theta$$

$$= \frac{1}{2.\pi}\frac{l_m^2}{2}\left[\left(\theta - \frac{1}{2}\sin(4.\pi)/2 - \sin(0)/2 = 0.0 = 0\sin(2\,\theta)\right)\right]_0^{2\pi}$$

$$= \frac{2.\pi}{2.\pi}\frac{l_m^2}{2} = \frac{l_m^2}{2}$$

$$= l_{RMS}^2\,\frac{l_m^2}{2} = \text{Therefore } l_{RMS} = \sqrt{\frac{l_m^2}{2}} = 0.707\,\text{l m}$$

Form Factor

The form factor for a sin wave is defined as the R.M.S value/Average value

$$\text{Form Factor of a sine wave} = \frac{\text{Root Mean Square value}}{\text{Average value}}$$

$$= \frac{0{,}707 \text{ Maximum value}}{0{,}637 \text{ Maximum value}} = 1.11$$

Peak Value

The peak or crest value of a sine wave is defined as the maximum value/the R.M.S value

$$\text{Peak/Creast value of a sine wave} = \frac{\text{Maximum value}}{\text{Root Mean Square value}}$$

$$= \frac{\text{Maximum value}}{0.707 \text{ Maximum value}} = 1.414$$

Vector Indication of AC values

A sinusoidal variable can be shown as a plot of amplitude against time or $\theta = \omega.t$

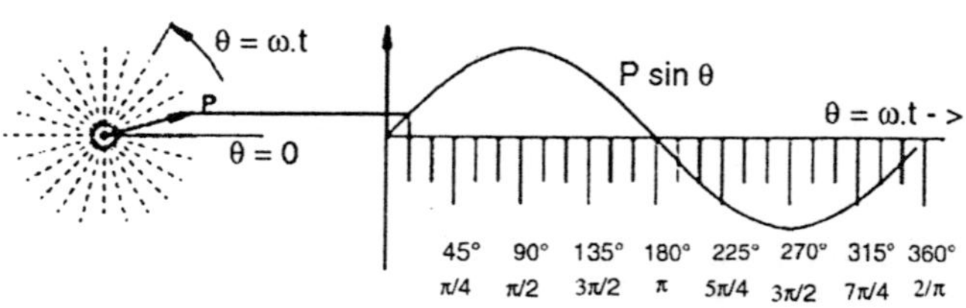

In can also be shown as a rotating vector or phaser. Both these methods are illustrated below. The rotating vector on the left is assumed to be rotating at a constant angular velocity ω Its projection on the vertical axis = P sin ω t. = P.sin θ is plotted as shown on the right

Two sinusoidal alternating variables with different amplitudes and phases are shown together as below.

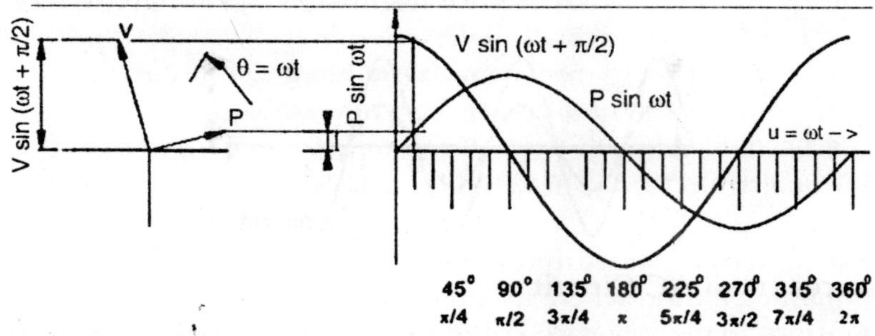

The two vectors can be added as shown below to obtain a resultant vector R.sin (ωt + z).

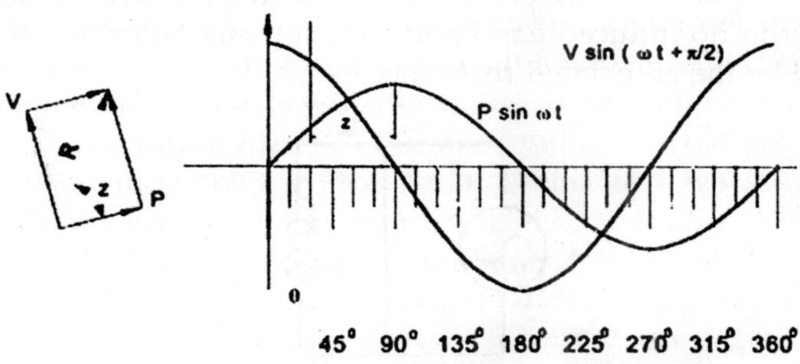

R and z are easily calculated as below as shown.

$$R = \sqrt{P^2 + V^2} \quad z = \tan^{-1} (V/P)$$

The methods of showing and adding sinusoidal variables is used when manipulating electrical currents currents flowing in circuits which include resistances, inductances, and capacitances

Resistance In an AC Circuit

Considering a AC circuit with a resistance of R.

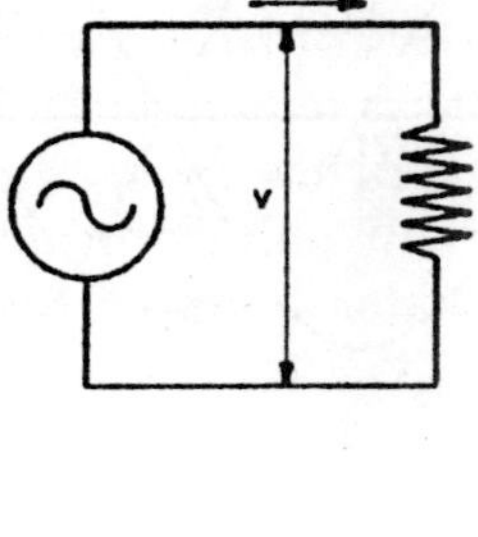

The ratio of the potential difference (V) across a resistor to the current (I) flowing through it is a constant - This is Ohms law. This applies to steady currents and to varying currents. For the case of an Alternating current the potential difference and the current are in phase and the current at any time =

$$i = v/R$$

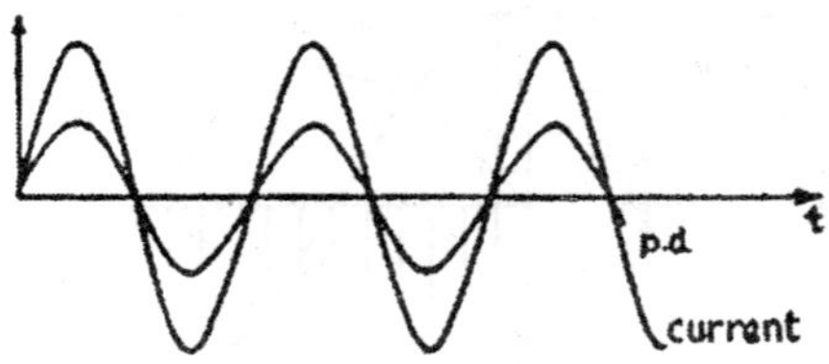

Inductances In an AC Circuit

Inductance (L) is measured using the unit of 1 henry which is the inductance in when 1 volt is produced by a current changing at the rate of 1 ampere/second. If there is a steady DC voltage across an inductance it will result in no induced e.m.f and the only effect limiting the steady current will be the associated resistance.

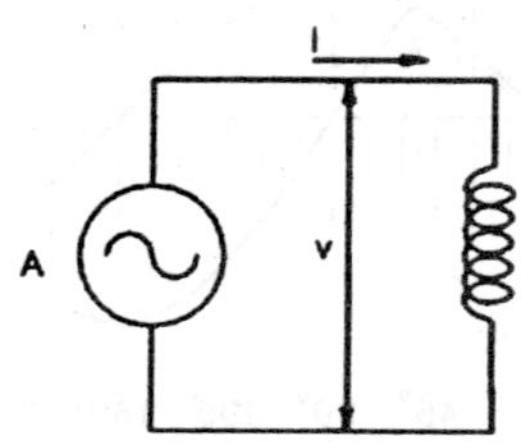

Consider a sine wave current through an inductance of L henries represented by the equation.. $i = I_m \sin \omega t$. (t being the time in second after the current has passed through zero from negative to positive.) The instantaneous value of the induced e.m.f resulting from this changing current =

$$e = -L\frac{di}{dt} = -L.I_m \frac{d \sin(\omega.t)}{dt} = -\omega.L.I_m \cos(\omega.t)$$
$$= \omega.\ L.I\ m \sin(\omega.t. - \pi/2)$$

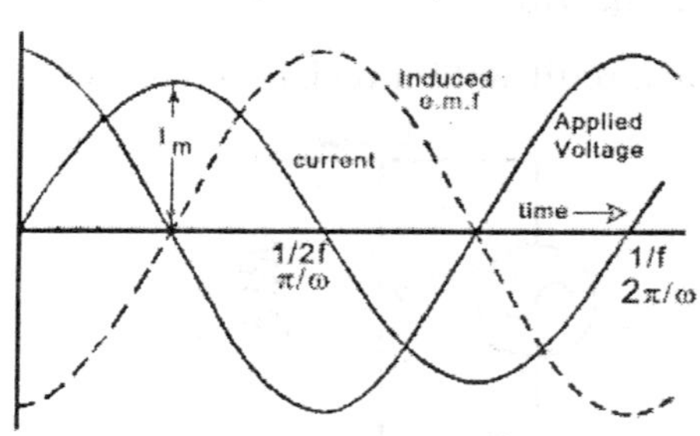

The amplitude of the induced e.m.f = ω.Λ.I μ and the resulting e.m.f sine wave effectively lags the current by ω/2. The sine waves are shown below

Note: The frequency $f = \omega/(2.\pi)$ and the wavelength = $1/f = 2.\pi/\omega$.

Assuming there is no resistance all of the applied voltage is absorbed in neutralising the induced e.m.f. Therefore

$$V = \omega.\Lambda.\lambda\ \mu \cos(\omega.t) = \omega.\Lambda.\lambda\ \mu \sin(\omega.t + \pi/2)$$

The applied voltage leads the current by π/2 and the maximum applied voltage = $\omega.L.I_m$ that is

$$V_m = \omega.L\ I\ m \text{ so that } \frac{V_m}{I_m} = \omega.L = 2.\pi.f.L$$

The inductive reactance is expressed as ohms and is represented by the symbol X_L. Therefore $I = V/X_L = V/(2.\pi.f.L)$. Unlike a resistance there is no energy absorbed in an inductance subject to a sine wave applied voltage.

Capacitances In an AC Circuit

The Capacitance C in which 1 coulomb is stored for a potential difference of 1 volt is a Farad

The charge on a capacitor Q = C.V and therefore dQ/dt = C dV/dt and therefore i = = C dV/dt

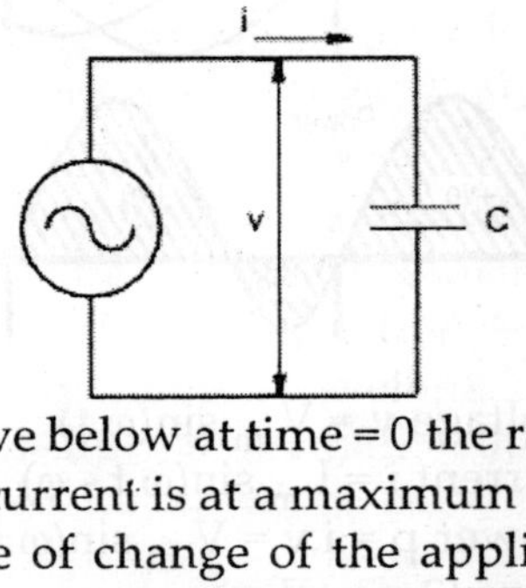

Referring to the curve below at time = 0 the rate of increase of applied voltage is such that the current is at a maximum value.In the period from t = 0 to t = π/2. The rate of change of the applied voltage reduces in a sinusoidal wave form to zero resulting in a fall of current flow to zero. The curve illustrates the relationship of the current to the applied voltage throughout the cycle

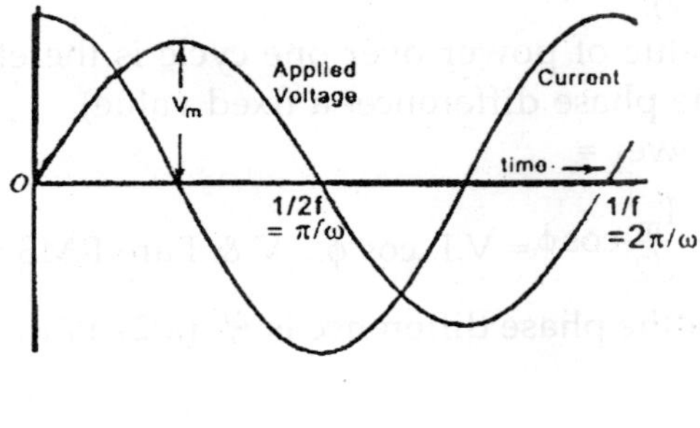

Note: The frequency f = ω/(2.π) and the wavelength = 1/f = 2.π/ω

Considering sinusoidal voltage applied across a purely capacitive circuit the following relationships apply.

$$V = V_m \sin\theta = V_m \sin\omega t$$

As the instantaneous value of the current is i = C.dv/dt then..

$$i = \frac{dv}{dt} = C.\frac{d}{dt}(V_m \sin\omega t) = \omega.C.V_m.\cos\omega t$$

$$= \omega.\ C.Vm.\ \sin(\omega t + \pi/2)$$

It is clear that the current leads the voltage by π/2 and the maximum value = $\omega C\ V_m$

$$X_C = \frac{V_m}{I_m} = \frac{V}{I} = \frac{1}{2\pi fC}$$

Power Factor

Consider a conductor with applied AC voltage with an out of phase current. The curve below shows an applied voltage with a current which is lagging the voltage by φ.

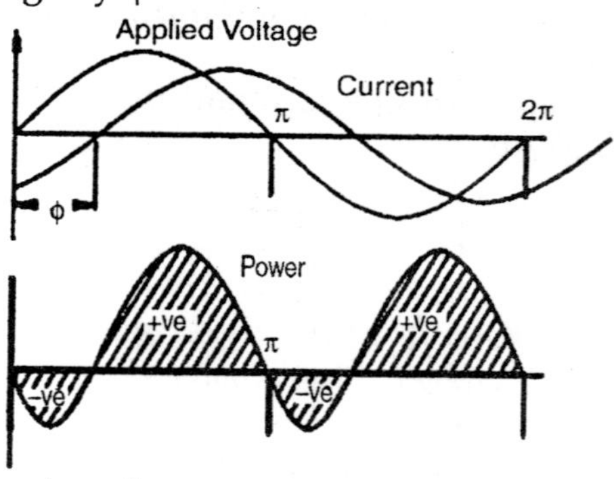

The instantaneous voltage v = V_m sin(ω.t)

The instantaneous current i = I_m sin(ω.t - φ)

The instantaneous power p = i.v = V_m sin(ω.t).I_m sin(ω.t - φ)

= 0,5.$V_m I_m$.[cos φ - cos (2ω - φ)]

= 0,5.$V_m I_m$.(cos φ) - 0,5.$V_m I_m$.cos (2ω - φ)

Over a complete cycle the average value of cos (2ω - φ) = 0. (It is varying sinusoidally)

The average value of power over one cycle is therefore. 0,5.$V_m I_{mx}$. (cos φ). (χοσ φ is the phase difference- a fixed value)

The average power =

$$Pav = \frac{V_m}{\sqrt{2}}\frac{I_m}{\sqrt{2}}\cos\phi = V.I.\ \cos\phi....V\ \&\ I \text{ are RMS values}$$

It is clear that if the phase difference is +/- (π/2) a purely capacitive or inductive load then the power from the alternator = 0. If the phase difference is 0 Then the power from the alternator is V. I.

The product of the RMS voltage and current (V.I) is the volt-amperes which is not the same as V.I. cos φ which is the power expressed in watts. The factor cos φ is known as the power factor and is an important factor in power distribution systems.

Power in watts = Voltamperes power factor = V.I.cos φ

Power factor (cos φ) = power (watts)/V.I

$= I^2 R/V.I = I R/V = IR/I.Z = R/Z$

Z = impedance of circuit.

For an alternative view on the topic consider an alternating voltage V with a current lagging by φ shown vectorially below.

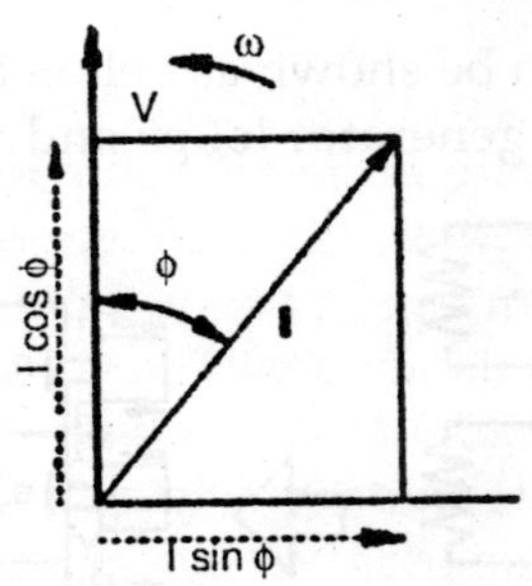

The current can be resolved into two components I cos φ in line with the voltage and I sin j at right angles to the voltage. The power due to the first component is VI cos j is the power or active component. The power due to the second component = V (I.sin j) cos 90^0 = 0 this is the reactive or wattless component of the current

Generation of Three Phase Supplies

Consider three loops each position 120° relative to each other rotating in a magnetic field as shown below:

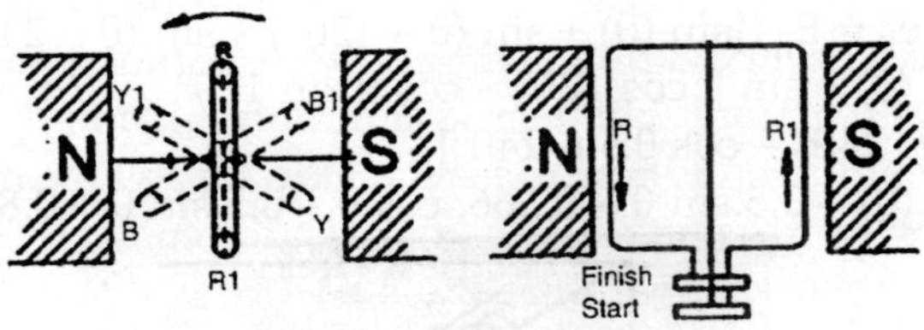

The loops are being rotated anti-clockwise and each loop is producing exactly the same emf with the same amplitude and frequency but the loop Y Lags loop R by 120^0 and the loop B lags loop Y by 120°. This is the sam e for the associated loop Y_1, B_1, and R_1. At any moment the e.m.f generated in the three loops are as follows

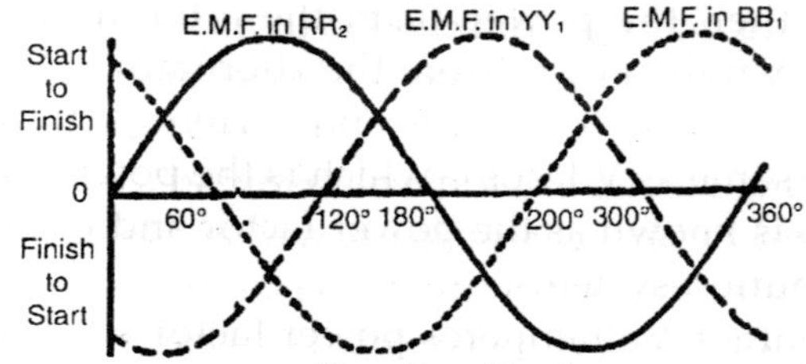

Three Phase Waveform

$$e_R = E_m \sin(\theta)$$
$$e_Y = E_m \sin(\theta - 120°)$$
$$e_B = E_m \sin(\theta - 240°)$$

Mesh/Delta Connection of Three Phase Supplies

The three phases can be shown as below as separate circuits with six conductors between the generator loops and the respective loads

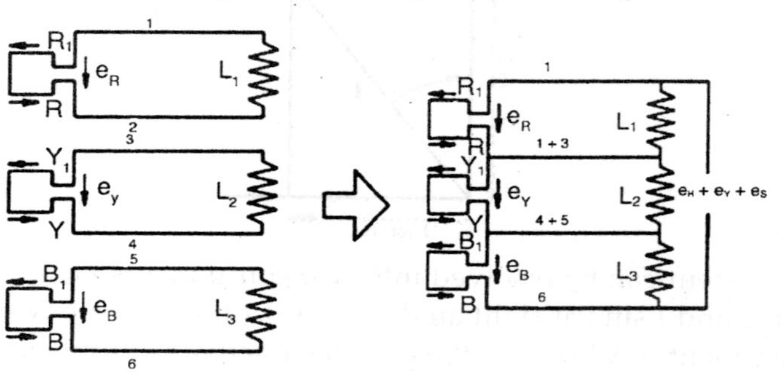

This system results in six conductors and can be simplified to reduce the number of conductors. Initially it is possible to replace conductors 2 and 3 with one conductor and conductors 4 and 5 with one conductor. It can also easily be proved that there is no voltage difference between 6 and 1 allowing these also to be connected as one conductor and any stage in the cycle, as follows

$$e_R + e_Y + e_B = E_m [\sin(\theta) + \sin(\theta - 120°) + \sin(\theta - 240°)]$$
$$= E_m [\sin(\theta) + \sin\theta.\cos 120° - \cos\theta.\sin 120°$$
$$+ \sin\theta.\cos 240° - \cos\theta.\sin 240^0]$$
$$= E_m [\sin(\theta) - 0{,}5.\sin\theta - 0{,}866.\cos\theta - 0{,}5.\sin\theta + 0{,}866.\cos\theta] = 0$$

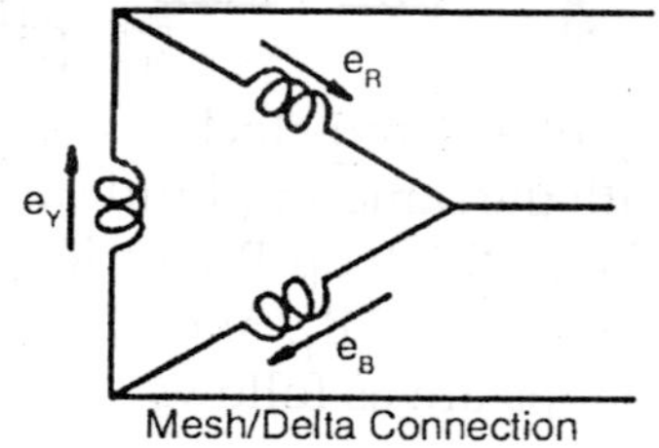

Mesh/Delta Connection

This method of connecting the conductors in a three-phase system is called a Mesh or Delta connection and is generally shown in figure.

Star Connection of Three Phase supplies

If the three start conductors 1,3 and 5 connected as one conductor NM as shown (in blue) below then the current flow is as shown. Three phase motors can be connected to the line conductors R.Y and B while lights, heaters etc are usually connected between the line conductors and the neutral line as shown by (L_1,L_2 & L_3)

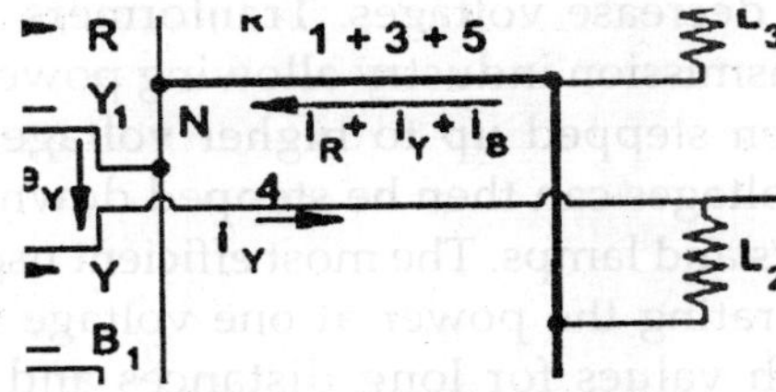

This conductor arrangement is normally shown as below

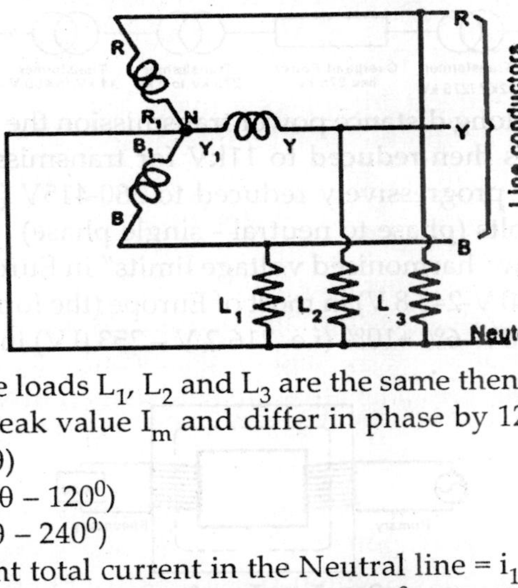

Ideally if the loads L_1, L_2 and L_3 are the same then the phase currents have the same peak value I_m and differ in phase by 120^0

$i_1 = I_m \sin(\theta)$

$i_2 = I_m \sin(\theta - 120^0)$

$i_3 = I_m \sin(\theta - 240^0)$

The resultant total current in the Neutral line = $i_1 + i_2 + i_3$

$= I_m[\sin(\theta) + \sin(\theta - 120^0) + \sin(\theta - 240^0] = I_m . (0) = 0$

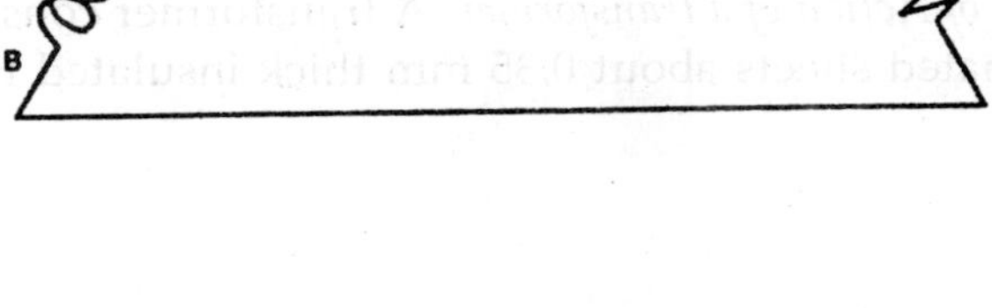

Hence with balanced loads there is no load in the neutral line at any instant and the neutral line can be removed resulting in the three wire star system as shown in figure. It can be proved that at every instant, for balanced loads the algebraic sum of the currents flowing in the three conductors is zero

Transformers

A transformer is an electromagnetic device having two or more stationary coils coupled through a mutual magnetic flux. Transformer are used to increase or decrease voltages. Tranformers are essential in the electrical power transmission industry allowing power to be generated at one voltage and then stepped up to higher voltages for long distance transmission. The voltages can then be stepped down for driving motors and operating heaters and lamps. The most efficient use of electrical power is achieved by generating the power at one voltage then increasing the voltage to very high values for long distances and then reducing the voltage to lower values for safe use in industry and in the home.

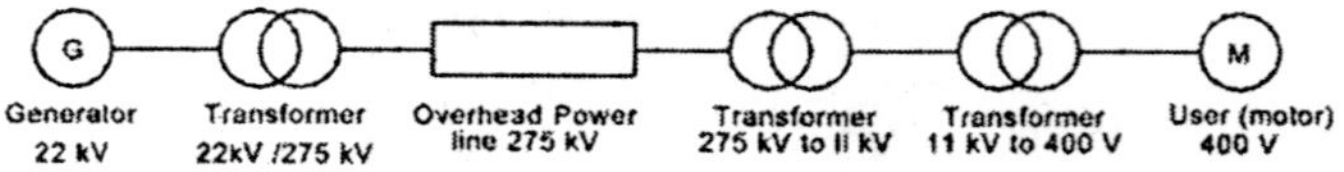

In the UK for long distance power transmission the voltage is stepped up to 275 kV. It is then reduced to 11kV for transmission within urban areas. It is locally progressively reduced to 380-415V (phase to phase -3 phase) 220-240 Volts (phase to neutral - single phase)

Note: The new "harmonised voltage limits" in Europe are now: 230V -10% +6% (i.e. 207.0 V-243.8 V) in most of Europe (the former 220V nominal countries), and 230V -6% +10% (i.e. 216.2 V – 253.0 V) in UK (former 240V nominal).

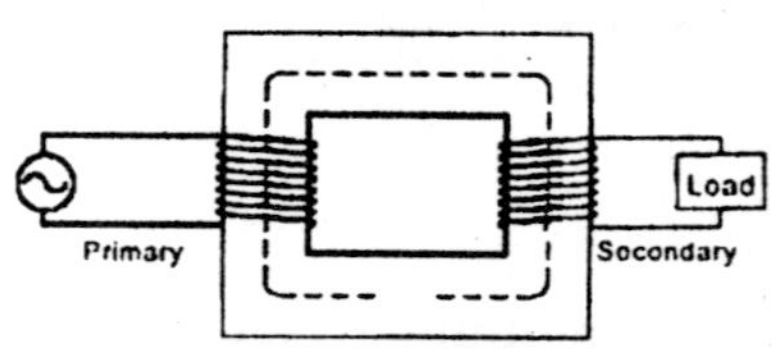

Core Type Transformer

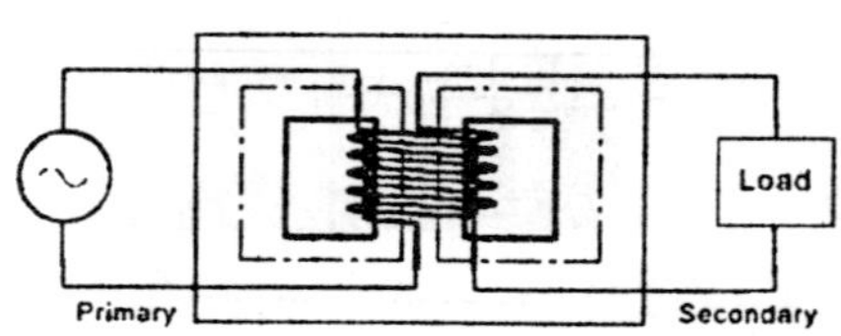

Shell Type Tranformer

Principle of Action of a transformer. A transformer consists of an metal core of laminated sheets about 0,35 mm thick insulated from each other to reduce eddy current losses. The vertical parts as shown are called the limbs and the horizontal parts are called the yokes. The two coils are wound on the limbs. One coil is called the primary coil and the other is called the secondary coil.

The metal core is made from a ferromagnetic metal selected because of its magnetic properties

An alternating voltage applied to the primary coil results in an alternating current which produces an alternating magnetic flux in the metal core. The current/flux level is just sufficient to induce an e.m.f that is practically equal and opposite to the applied voltage. The mean path of the alternating flux is indicated in the figure. Assuming all of the flux passes through the primary and the secondary winding, it induces the same e.m.f. in each turn of both coils. Hence if N_1 and N_2 are the number of turns in the primary and secondary coils respectively.

$$\frac{\text{e.m.f inducedin sec ondary winding}}{\text{e.m.f induced inprimary winding}} = \frac{N_2.\text{e.m.f / turn}}{N_1.\text{e.m.f / turn}} = \frac{N_2.}{N_2}$$

When the secondary winding is in open circuit its terminal voltage is the same as the induced e.m.f. The primary current is such that the induced e.m.f is equal and opposite to the applied alternating voltage.

When a load is connected to the secondary winding a current flows. This secondary current induces a magnetic flux which tends to reduce the current flow (Lenz's Law). The induced flux tends to reduce the magnetic flux and the resulting back back e.m.f. in the primary windigs This changes the difference between the applied voltage and the back e.m.f. in the primary windings from (say 0,05% to 1%).

As a result the primary current increases, by say 20 times the no-load current.

The demagnetising ampere turns of the secondary winding are thus neutralised by the increase in the primary ampere turns.. As the initial no-load primary ampere-turns are very small it can be stated that

Full load Ampere-Turns on (Primary) = Full load Ampere-Turns on (Secondary)

$$I_1.N_1 = I2.N2. \text{ Therefore } I_1/I_2 = N_2/N_1 = V_2/V_1$$

In practice the above are all approximate *Transformer Losses*. The energy wasted in a transformer can be divided into magnetic leakage losses copper losses and iron losses.

The iron losses are hysteresis losses and eddy current losses. The hysterisis losses are described on webpage *Electromagnetism - Hysteresis Losses*.

Magnetic leakage occurs because all of the flux established by the current in the primary coil does not link with the secondary and all of the flux due to the current induced in the secondary coil does not link with the primary coil. The effect of this is to produce a slightly lower secondary voltage than that calculated by the relevant formula. In commercial transformers there is virtually zero magnetic losses.

The copper losses are simply the normal losses to heat as a result of the resistance to of the wires making up the primary and secondary windings. These losses are related to I^2 R and therefore increase with the power transferred by the transformer. The copper power losses are simply the sum of the I^2.R of each winding.

The eddy current losses are caused by induced currents circulating in the transformer core and, to a small extent, in the copper windings. If the core were solid then the induced current would be significant because there would be very low resistance. By making the core from thin laminated steel plates separated by insulation the conduction path is reduced such that the eddy current losses are minimised. The effect of these losses is to reduce the secondary voltage below that obtained by the relevant formula. Refer to figure below..

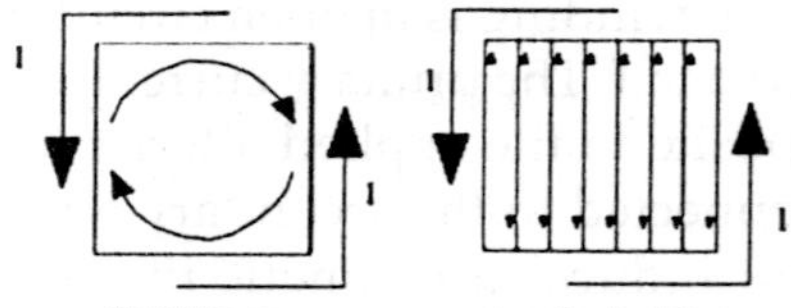

Solid Core Laminated Core

Motors: There are a number of methods of classifying AC motors

1. Single Phase and Three phase.. Most AC motors are three phase.
2. Motors with shunt characteristics relating to torque and speed and motors with series type characteristics and motors running at constant speed (synchronous speed)
3. Motors in which all of the current is supplied from outside and motors in which the current in one part (generally the rotor) are produced by induction.

Single phase motors with series characteristics are used primarily for fraction kW motors used for vacuum cleaners, small tools etc. Single phase induction motors are only self starting if special provisions are provided to drive them up to synchronous speed.

The vast majority of motors are polyphase induction motors. An induction motor consists of a rotor and a stator. For notes on AC inductance motors ref to webpage *AC Motors*

A synchronous motor is basically an alternator which is brought up to a synchronous speed and then connected to an external polyphase supply.

It will then run as a motor at constant speed. Synchronous motors are not self starting and they require a source of direct current (usually an exciter on the motor shaft) to provide the exciting field on the rotor. Synchronous motors will, if over excited run at a low leading power factor which will compensate for the low lagging power of other loads such as induction motors.

Synchronous motors are used in industry to compensate for the lagging power factors associated with induction motors providing a higher overall power factor. Synchronous motors are sometimes used when a long term constant speed at load is required with infrequent starts

Chapter 2

Thermoelectricity

INTRODUCTION TO THERMOELECTRICITY

The 3 TE effects are named for Seebeck, Peltier, and Thomson. The easiest way to think about thermoelectricity is to realise that electrical and thermal currents are coupled. The particles that carry electric charge in a material also carry heat. The three TE effects are just three ways in which this coupling is realized, and all are closely related. In fact, knowing the Seebeck coefficient as a function of temperature allows one to calculate the Peltier and Thomson coefficients.

Below are some cartoons and descriptions that should be helpful in understanding the different effects and how they are related.

(1) Seebeck Effect

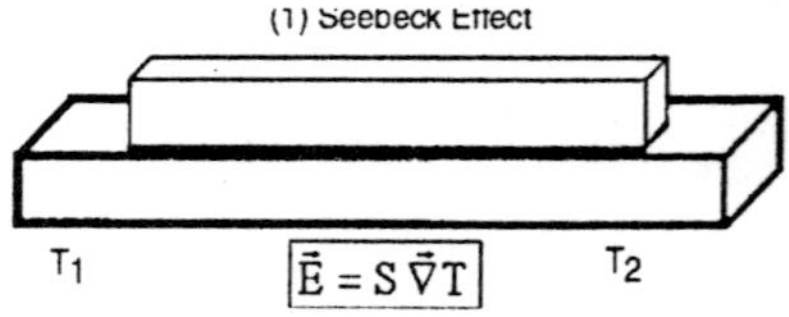

Imagine a conductor with an applied temperature gradient. Electrons (or holes) will thermally diffuse from the hot end to the cold end, and carry their charge with them. This charge builds up on the cold end and creates an electric field inside the sample. When the current generated by this electric field cancels the flow due to thermal diffusion, equilibrium is reached. The Seebeck coefficient (also called the thermopower) is the ratio between the electric field and the temperature gradient (or equivalently, the ratio between the voltage difference and temperature difference between the ends of the sample).

So, *roughly speaking,* the Seebeck coefficient can be thought of as a measure of the coupling between the thermal and electrical currents in a material. Note that the sign of the Seebeck coefficient depends on the sign of the charge carriers (whether positive or negative charge builds up on the cold end). So measuring the Seebeck coefficient is a way to determine whether the charge carriers in a particular material are holes or electrons.

NOTE: What we have described here is known as the absolute

Seebeck effect. If we wanted to measure this voltage, the wires that the attach to the ends of the sample to hook up to our voltmeter would also develop Seebeck voltages. What we would measure is the difference between our wires and our sample. This is called the differential Seebeck effect. See the discussion of thermocouples below.

(2) Peltier Effect

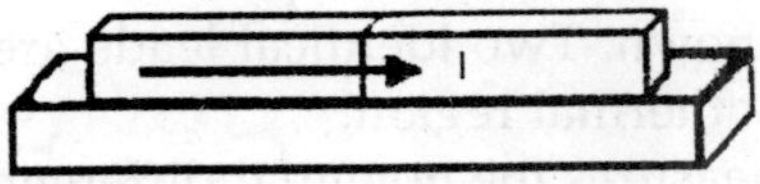

$\Pi_{12}\, I \equiv$ Power evolved at junction

Now consider passing a current through a junction between two different materials held at a constant temperature. This electrical current will drag along a thermal current, the magnitude and direction of which depends on the Seebeck coefficients. If they are different, thermal energy will be leaving the junction at a different rate than it is entering.

Thus heat is generated or absorbed at the junction. Imagine that the material on the left is p-type (positive charge carriers moving to the right) and the one on the right is n-type (negative charge carriers moving to the left). Both materials will be carrying heat towards the junction, and power will be evolved there. The Peltier coefficient of the junction is a property depending on both materials and is the ratio of the power evolved at the junction to the current flowing through it.

(3) Thomson Effect

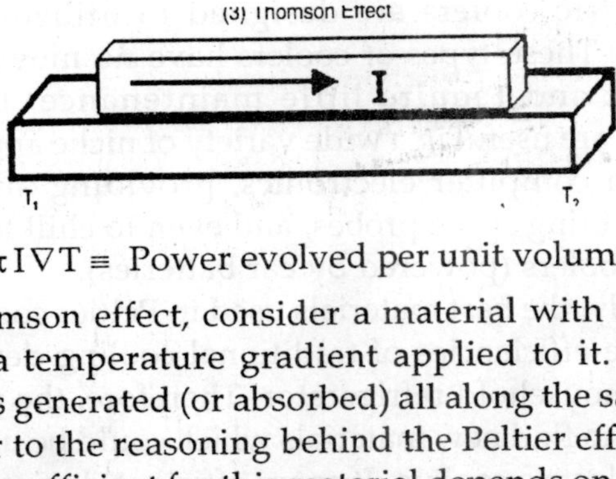

$\tau I \nabla T \equiv$ Power evolved per unit volume

For the Thomson effect, consider a material with a current flowing through it and a temperature gradient applied to it. In this situation, thermal energy is generated (or absorbed) all along the sample. The reason for this is similar to the reasoning behind the Peltier effect, if one realizes that the Seebeck coefficient for this material depends on temperature. This means that the Seebeck coefficient is different at different places along the sample. So the sample can be thought of as a series of many small Peltier junctions, each of which is generating (or absorbing) heat. The Thomson coefficient is the ratio of the Power evolved per unit volume in the sample to the applied current and temperature gradient.

Thermocouples are the most common examples of TE devices. They take advantage of the differential Seebeck effect. A diagram of a thermocouple is shown figure.

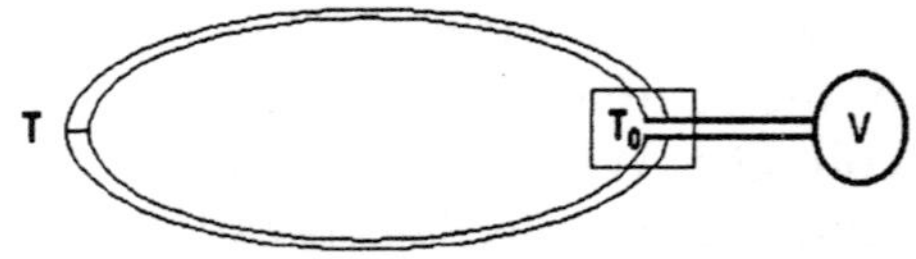

The two legs of the circuit are made of different materials, and the two junctions are held at different temperatures. T is the temperature to be measured, T_0 is known. Two identical leads are connected to a break in the circuit in an isothermal region.

The voltmeter measures the potential difference between the ends of the leads. Since the leads are identical and the temperature difference between the ends of the leads is the same, their contribution to the voltage cancels.

What is measured is the difference between the thermoelectric voltage generated by the two legs of the thermocouple. If the Seebeck coefficients are know, this gives the difference between the temperatures T and T_0.

THERMOELECTRIC MATERIALS

Thermoelectric materials take advantage of the coupling between thermal and electrical currents, and are used for the direct conversion between thermal and electrical energy. With these materials, electricity can be used to pump heat (thermoelectric coolers) or waste heat can be used to generate electricity (thermoelectric generators).

Thermoelectric coolers are designed to utilize the Peltier effect described above. These types of coolers have no moving parts and as a result are quiet and require little maintenance. Because of these advantages, they are useful in a wide variety of niche applications: cooling laser diodes and computer electronics, providing air conditioning in submarines, powering space probes, and even to chill food and drinks in portable picnic coolers (powered by car batteries).

Unfortunately, the best materials used in Peltier coolers today cannot compete with the efficiencies of traditional cooling devices, such as the compressor in a household refrigerator. Therefore, the primary objective of our research is to find new materials which could be used to make more efficient thermoelectric coolers. If more efficient devices could be made, there would be a number of new and exciting applications for Peltier coolers.

The Peltier junction mentioned above consists of two materials, one with a positive thermopower and one with a negative thermopower. Since the charge carriers in p-type materials and n-type materials have opposite sign, their thermopowers have opposite sign.

Schematic of a TE device. Since the charge carriers move in a different direction in each leg, each carries heat away from the cold end.

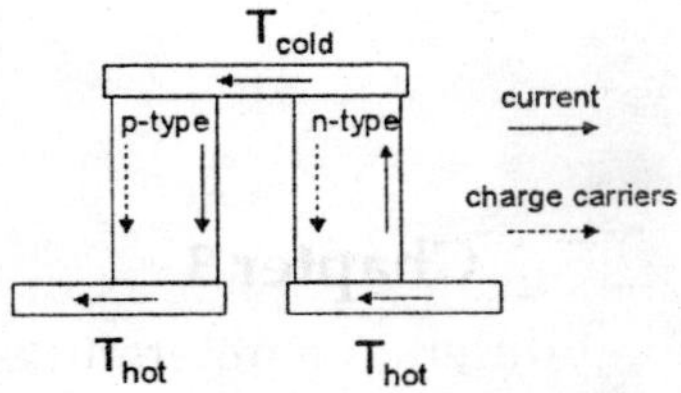

It is more than just the thermo power that determines the performance of a device. We need the electrical resistivity to be small so energy is not wasted in Joule heating.

We also need the thermal conductivity to be small so heat we pump to the hot end stays there. For maximum device efficiency, we need to maximize each material's dimensionless Figure of Merit (ZT).

$$ZT \frac{TS^2}{\rho k}$$

S = thermopower (V/K)

p = electrical resistivity (Wcm)

k = thermal conductivity (W/cmK)

T = temperature (K)

At present, the best materials are small band gap semiconductors; optimized Bi_2Te_3 has S = 220 μV/K and ZT = 1 at room temperature. However, a ZT greater than 3 is needed to compete with traditional cooling technologies. Our goal is to find new materials with ZT > 1 at or below room temperature.

Chapter 3

Voltaic Electricity

VOLTAIC OR GALVANIC ELECTRICITY

The second method of generating *electricity* is by chemical means, so called, because a liquid is used as one of the agents.

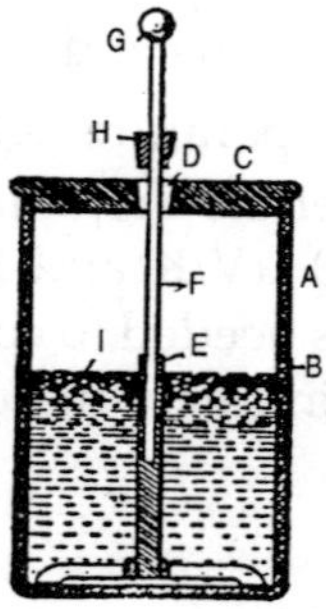

Fig. Leyden Jar

Galvani, in 1790, made the experiments which led to the generation of electricity by means of liquids and *metals*. The first *battery* was called the "crown of cups," shown in figure, and consisting of a row of glass cups:

(a) Containing salt water. These cups were electrically connected by means of bent metal strips.
(b) Each strip having at one end a copper *plate*.
(c) And at the other end a zinc plate.

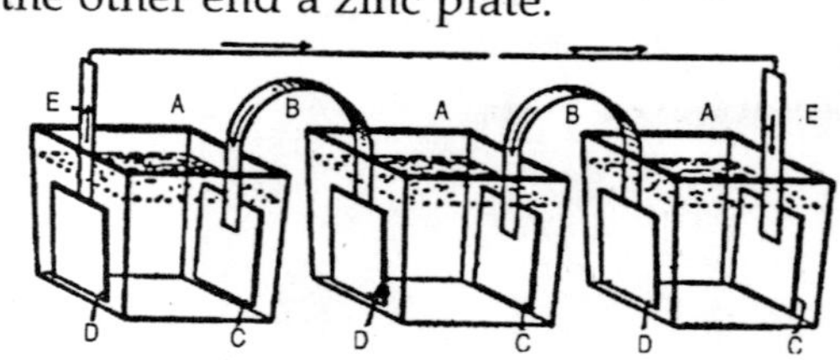

Fig. Galvanic Electricity. Crown of Cups.

(d) The first plate in the cup at one end is connected with the last plate in the cup at the other end by a *conductor*.

(e) To make a complete circuit.

THE CELL AND BATTERY

From the foregoing it will be seen that within each cup the *current* flows from the zinc to the copper plates, and exteriorly from the copper to the zinc plates through the conductors (B and E). A few years afterwards Volta devised what is known as the voltaic pile.

VOLTAIC PILE - HOW MADE

This is made of alternate discs of copper and zinc with a piece of cardboard of corresponding size between each zinc and copper plate. The cardboard discs are moistened with acidulated water. The bottom disc of copper has a strip which connects with a cup of *acid,* and one wire *terminal* (A) runs there from.

The upper disc, which is of zinc, is also connected, by a strip, with a cup of acid from which extends the other terminal wire (B).

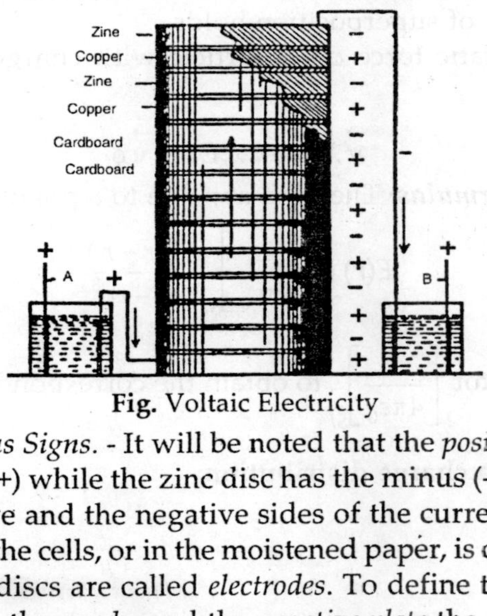

Fig. Voltaic Electricity

Plus and Minus Signs. - It will be noted that the *positive* or copper disc has the plus sign (+) while the zinc disc has the minus (-) sign. These signs denote the positive and the negative sides of the current.

The liquid in the cells, or in the moistened paper, is called the *electrolyte* and the plates or discs are called *electrodes.* To define them more clearly, the *positive plate* is the *anode,* and the *negative plate* the *cathode.*

The current, upon entering the zinc plate, decomposes the water in the electrolyte, thereby forming *oxygen.* The hydrogen in the water, which has also been formed by the decomposition, is carried to the copper plate, so that the plate finally is so coated with hydrogen that it is difficult for the current to pass through. This condition is called "polarization," and to prevent it has been the aim of all inventors.

Chapter 4

Electrodynamics

ELECTROSTATICS

The fundamental equations of electrostatics are linear equations,

$\vec{\nabla}.E = \frac{\rho}{\varepsilon_0}$, $\vec{\nabla} \times E = 0$ (SI units), $\vec{\nabla}.E = 4\pi\rho$, $\vec{\nabla} \times E = 0$ (Gaussian units).

The principle of superposition holds.

The electrostatic force on a particle with charge q at position r is $F = qE(r)$.

$$\vec{\nabla}.E = 0 \Leftrightarrow E = -\vec{\nabla}\phi$$

Important formulas: The field at r due to a point charge at r':

$$E(r) = \left[\frac{1}{4\pi\varepsilon_0}\right]_{SI} \frac{q(r-r')}{|r-r'|^3}$$

(Omit the factor $\left[\frac{1}{4\pi\varepsilon_0}\right]_{SI}$ to obtain the corresponding expressions in Gaussian units.)

The field of a charge distribution:

$$E(r) = \left[\frac{1}{4\pi\varepsilon_0}\right]_{SI}$$

$$\left[\int_{V'} \frac{\rho(r')(r-r')}{|r-r'|^3} dV' + \int_{V'} \frac{\sigma(r')(r-r')}{|r-r'|^3} dA' + \int_{A'} \frac{\lambda(r')(r-r')}{|r-r'|^3} dl' + \sum_i \frac{q_i(r')(r-r_i)}{|r-r_i|^3}\right]$$

(We consider volume, surface, and line charge distributions and point charges.)

The potential at *r* due to a point charge at *r'*:

$$\phi(r) = \left[\frac{1}{4\pi\varepsilon_0}\right]_{SI} \frac{q}{|r-r'|}$$

The potential of a charge distribution:

$$f(r) = \left[\frac{1}{4\pi\varepsilon_0}\right]_{SI}$$

$$\left[\int_{V'} \frac{\rho(r')}{|r-r'|} dV' + \int_{A'} \frac{\sigma(r')}{|r-r'|} dA' + \int_{l'} \frac{\lambda(r')}{|r-r'|} dl' + \sum_i \frac{q_i}{|r-r_i|}\right]$$

Gauss' law:

$$\oint_A E.\hat{n}\, dA = \frac{Q_{inside}}{\varepsilon_0} \text{ (SI units)}, \oint_A E.\hat{n}\, dA = 4\pi\, Q_{inside} \text{ (Gaussian units)}.$$

In situations with enough symmetry, Gauss' law alone can be used to find E.

The electrostatic energy of a charge distribution:

$$U = \frac{1}{2}\int \phi(r')dq(r') = \frac{1}{2}$$

$$\left[\int_{V'} \rho(r')\phi(r')dV' + \int_{A'} \sigma(r')\phi(r')dA + \int_{l'} \lambda(r')\phi(r')dl' + \sum_i q_i\phi(r_i)\right]$$

or, for a continuous charge distribution,

$$U = \frac{\varepsilon_0}{2}\int_{dl\_space} E.E.dV \text{ (SI units)}, U = \frac{1}{8\pi}\int_{dl\_space} E.E.dV \text{ (Gaussian units)}.$$

DIPOLES

The field of a dipole at the origin:

$$E(r) = \left[\frac{1}{4\pi\varepsilon_0}\right]_{SI} \frac{1}{r^3}\left[3(p.\hat{r})\hat{r} - p\right]$$

The potential of a dipole at the origin:

$$\phi(r) = \left[\frac{1}{4\pi\varepsilon_0}\right]_{SI} \frac{p.\hat{r}}{r^2}$$

The force on a dipole:

$$F = \vec{v}(p.E)$$

The torque on a dipole:

$$\tau = -p.E$$

The energy of a dipole in an external field:

$$U = -p.E$$

Properties of conductors in electrostatics

$E = 0$ inside.

Any excess charge resides on the surface.

E on the surface is perpendicular to the surface.

$E = \frac{\sigma}{\varepsilon_0}\hat{n}$ (SI units), $E = 4\pi\sigma\hat{n}$ (Gaussian units), just outside the surface.

DIELECTRICS

The polarization $P = \frac{dp}{dV}$ is defined as the dipole moment per unit volume.

The total charge density is due to free and to bound (polarization) charges.

$\rho = \rho_l + \rho_p,\ \sigma = \sigma_f + \sigma_p,\ \rho_p = -\vec{v}.P,\ \sigma_p = P.\hat{n}$

$D = \varepsilon_0 E + P$ (SI units), $D = E + 4\pi P$ (Gaussian units).

For linear, isotropic, homogeneous dielectrics we have

$$P = \varepsilon_0 \chi e E,\quad P = \chi e E,$$

$$D = \varepsilon_0(1 + \chi e)\, E = \varepsilon_0 k_e E = \varepsilon E,\quad D = (1 + 4\pi\chi e)\, E = \varepsilon E,$$

$$\vec{v}.D = \rho t,\ v^2\phi = \frac{\rho f}{\varepsilon},\ \vec{v}.D = 4\pi\rho f,\ v^2\phi = -4\pi\frac{\rho f}{\varepsilon}$$

Boundary Conditions in Electrostatics

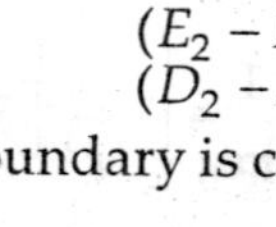

$(E_2 - E_1).\hat{n}_2 = \frac{\sigma}{\varepsilon_0}$ (SI units),	$(E_2 - E_1).\hat{n}_2 = 4\pi\sigma$ (Gaussian units)
$(E_2 - E_1).\hat{t} = 0,$	$(E_2 - E_1).\hat{t} = 0$
$(D_2 - D_1).\hat{n}_2 = \sigma_f,$	$(D_2 - D_1).\hat{n}_2 = 4\pi\sigma_f,$
is continuous across the boundary	is continuous across the boundary.

METHODS FOR SOLVING ELECTROSTATIC PROBLEMS

The Multipole Expansion

For a charge distribution located near the origin we have

$$f(r) = \left[\frac{1}{4\pi\varepsilon_0}\right]_{SI}$$

$$\left[\frac{1}{r}\int_{V'}\rho(r')\,dV' + \frac{1}{r^2}\hat{r}.\int_{V'}\rho(r')r'\,dV' + \frac{1}{r^3}\int_{V'}\frac{3(\hat{r}.r')^2 - r'^2}{2}\rho(r')dV' + \ldots\right]$$

with

$$Q_{ij} = \int \left(x_i' x_j' - \frac{1}{3}\delta_{ij} r'^2 \right) \rho(r') dV', \; Q_{ij} = Q_{ij}, \; \sum_i Q_{ii} = 0$$

Here Q = total charge, p = dipole moment, Q_{ij} = quadrupole moment tensor of the charge distribution. If the problem has rotational symmetry about the z-axis, such that $Q_{xx}=Q_{yy}=-\frac{1}{2}Q_{zz}$, then Q_{zz} is called the quadrupole moment.

In spherical coordinates we have

$$\phi(r) = \left[\frac{1}{4\pi\varepsilon_0}\right]_{SI} \sum_{l=0}^{\infty} \sum_{m=-l}^{l} \frac{4\pi}{2l+1} q_{lm} \frac{Y_{lm}(\theta,\phi)}{r^{l+1}}$$

with

$$q_{lm} = \int Y_{lm}^{\#}(\theta',\phi')\rho(r')(r')dV'$$

$$q_{l-m} = (-1)^m q_{lm}$$

The q_{lm} are called the multipole moments.

Relationship between Cartesian and spherical multipoles:

$$q_{00} = \frac{1}{\sqrt{4\pi}} q \quad q_{20} = \frac{1}{2}\sqrt{\frac{5}{4\pi}} Q_{33}$$

$$q_{10} = \sqrt{\frac{3}{4\pi}} p_z \quad q_{21} = -\frac{1}{3}\sqrt{\frac{15}{18\pi}} (Q_{33} - iQ_{23})$$

$$q_{11} = \sqrt{\frac{3}{8\pi}} (p_x - ip_y) \quad q_{22} = \frac{1}{12}\sqrt{\frac{15}{2\pi}} (Q_{11} - 2iQ_{12} - Q_{22})$$

The energy of a charge distribution near the origin in an external field is given by

$$W = q\phi(0) - p.\, E(0) - \frac{1}{2}\sum_i \sum_j Q_{ij} \frac{\partial E_j}{\partial x_i}\Big|_{x_i=0} + \cdots$$

This expansion shows how the various multipoles interact with the external field.

The Method of Images

Consider a charge q at $d = d\hat{z}$.

Assume the $z = 0$ plane is a grounded conducting plane. Then placing an image charge

$$q' = -q\frac{R}{d} \text{ at } d' = \frac{R^2}{d}$$

on the z-axis makes the $z = 0$ plane an equipotential with = 0.

Assume a grounded conducting sphere of radius R is centered on the origin. Then placing an image charge on the z-axis makes the sphere an equipotential with □ = 0. Add q'' at the centre of the sphere to change the potential or the total charge on the sphere.

Consider a line charge parallel to the x-axis at $d = d\hat{z}$.

Assume a conducting cylinder of radius R has as its axis the x-axis. Then placing an image charge

$$\lambda' = -\lambda \text{ at } d'\hat{z} \text{ with } d'\frac{R^2}{d}$$

parallel to the x-axis makes the sphere an equipotential with

$$\phi = \frac{\lambda}{2\pi\varepsilon_0}\ln\frac{R}{d}.$$

Add V_0 everywhere to change the potential of the cylinder.

Assume the $z = 0$ plane is a plane interface between two dielectrics.

Consider a charge q at $d = d\hat{z}$ in $_1$. Then placing an image charge

$$q' = -q\frac{\varepsilon_2 - \varepsilon_1}{\varepsilon_2 + \varepsilon_1}$$

at on the z-axis in a medium with $□_1$ gives the potential and field in dielectric 1. And placing an image charge at $d'' = d$ on the z-axis in a medium with $□_2$ gives the potential and field in dielectric.

Boundary Value Problems

Let $\phi = \phi$ (r, θ), $(0 \le \theta \le \pi)$ in spherical coordinates. Then

$$\phi(r, \theta) = \sum_{n=0}^{\infty}(A_n r^n + B_n r^{-(n+1)})P_n(\cos\theta)$$

Let $\phi = \phi$ (r, θ), $(0 \le \theta \le \pi)$ in cylindrical coordinates. Then

$$\phi(\rho,\phi) = \sum_{n=0}^{\infty}(A_n \cos n\phi + B_n \sin n\phi)\,\rho^n + \sum_{n=0}^{\infty}(A_n' \cos n\,\phi + B_n' \sin n\,\phi)$$

$\rho^{-n} + a_0 + b_0 \ln \rho$.

When solving electrostatic problems, we often rely on the uniqueness theorem. If □ or its normal derivatives are specified at the boundaries of a volume V, then a unique solution exists for □ inside V.

Useful Mathematical Formulas

$$\vec{\nabla}\frac{1}{|r-r'|} = -\frac{r-r'}{|r-r'|^3}, \nabla^2\frac{1}{|r-r'|} = -4\pi\delta(r-r')$$

$$\frac{1}{|r-r'|} = \sum_{n=0}^{\infty}\frac{r^n}{r_>^{n+1}}P_n(\cos y) = 4\pi\sum_{n=0}^{\infty}\frac{1}{2n+1}\left(\sum_{m=-n}^{n} Y_{nm}^{\theta}(\theta',\phi')Y_{nm}(\theta,\phi)\right)\frac{r^n}{r_>^{n+1}}$$

$$Y_{lm}(\text{q, f}) = \sqrt{\frac{2(l+1)(l-m)!}{4\pi(l+m)!}} P_l^m(\cos\theta)e^{im\phi}$$

$$P_0(x) = 1$$
$$P_1(x) = x$$

$$P_2(x) = \frac{3x^2 - 1}{2}$$

MAGNETOSTATICS

The fundamental equations magnetostatics are linear equations,

$$\vec{\nabla}.B = 0, \vec{\nabla} \times B = \mu_0 J = \frac{J}{\varepsilon_0 c^2}$$

$$\text{(SI units)}\ \vec{\nabla} B = 0, \vec{\nabla} \times B = \frac{4\pi}{c} J\ \text{(Gaussian units)}$$

The principle of superposition holds.
The magnetostaticstatic force on a particle with charge q is

$$F = qv \times B\ \text{(SI units)},$$

$$F = q\frac{v}{C} \times B\ \text{(Gaussian units)}.$$

Definitions:

Drift velocity: $< v > = \frac{1}{N}\sum_i v_i,$ = number of charge carriers

Current density: $n = nq < v > n = \frac{N}{V}$

Current: $I = \frac{dQ}{dt} = \int_A j.\hat{n}dA$

The continuity equation is $\vec{\nabla}.j = -\frac{\partial\rho}{\partial t}$. In statics $-\frac{\partial\rho}{\partial t} = 0 \Rightarrow \vec{\nabla}.\text{j} = 0$.

Ampere's Law

$$\oint_\Gamma B.dr = \mu_0 l_{through\_\Gamma}\ \text{(SI units)}$$

$$\oint_\Gamma B.dr = \frac{4\pi}{c} l_{through\_\Gamma}\ \text{(Gaussian units)}$$

In situations with enough symmetry Ampere's law alone can be used to find the magnitude of B. The flux of B through any closed surface is zero.

$$\int_A B.\hat{n}dA = 0$$

The Biot-Savart Law

(SI units)

$$B(r) = \frac{\mu_0}{4\pi}\int_{V'}\frac{j(r')\times(r-r')}{|r-r'|^3}dV'$$

$$B(r) = \frac{\mu_0}{4\pi}\int_{V}\frac{Idl'\times(r-r')}{|r-r'|^3}$$

(Gaussian units)

$$B(r) = \frac{1}{C}\int_{V'}\frac{j(r')\times(r-r')}{|r-r'|^3}dV'$$

$$B(r) = \frac{1}{C}\int_{V}\frac{Idl'\times(r-r')}{|r-r'|^3}$$

For filamentary currents we have

The Magnetic Vector Potential

$$\vec{v}.B = 0, \Leftrightarrow \mathrm{B} = \vec{v}\times 4$$

A is not unique. $A' = A + \vec{v}\Psi + C,$ with an arbitrary scalar field and C an arbitrary constant vector is also a vector potential for the same field.

In magnetostatics we choose $\vec{v}.A = 0$. Then

(SI units)

$$\nabla^2 A(r) = -\mu_0 j(r)$$

$$A(r) = \frac{\mu_0}{4\pi}\int_{V'}\frac{J(r')}{|r-r'|}dV'$$

Gaussian units)

$$\nabla^2 A(r) = -\frac{4\pi}{C}j(r)$$

$$A(r) = \frac{1}{C}\int_{V'}\frac{j(r')}{|r-r'|}dV'$$

The uniqueness theorem: If A or its normal derivatives are specified at the boundaries of a volume V, then a unique solution exists for A inside V.

Boundary Conditions in Magnetostatics

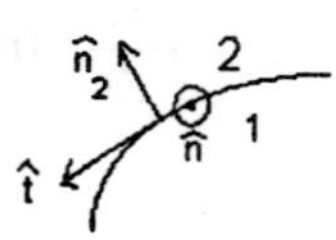

(SI units)

$$(B_2 - B_1).\ \hat{n}_2 = 0$$

$$(B_2 - B_1).\ \hat{t} = \mu_0 k.\hat{n}$$

□ is continuous across the boundary.

(Gaussian units)

$$(B_2 - B_1).\ \hat{n}_2 = 0$$

$$(B_2 - B_1).\ \hat{t} = \frac{4\pi}{C}k.\hat{n}$$

□ is continuous across the boundary.

The Force on a Current Distribution

(SI units) (Gaussian units)

$$F = \int_V j(r) \times B(r) dV \qquad F = \frac{1}{C}\int_V j(r) \times B(r) dV$$

For filamentary currents we have

$$F = \int_V Idl \times B(r) \qquad F = \frac{1}{C}\int_V Idl \times B(r)$$

The magnetic dipole moment of a charge distribution

(SI units) (Gaussian units)

$$m = m = IA\hat{n} = \frac{1}{2}\int_V r \times j(r) dV \qquad m = m = \frac{1}{2c}\int_V r \times j(r) dV$$

The vector potential of a magnetic dipole at the origin is

$$A(r) = \frac{\mu_0}{4\pi}\frac{m \times r}{r^2}, \qquad A(r) = \frac{m \times r}{r^2},$$

The magnetic field of a magnetic dipole at the origin is

$$B(r) = \frac{\mu_0}{4\pi}\frac{3(m.\hat{r})\hat{r} - m}{r^3}, \qquad B(r) = \frac{3(m.\hat{r})\hat{r} - m}{r^3},$$

The energy of a magnetic dipole in an external magnetic field is $U_{mech} = -m.B$. This is the mechanical work done to bring the dipole from infinity to its present position.

The force on a dipole is $F = \vec{\nabla}(m.B)$.

The torque on a dipole is $\tau = m \times B$.

Magnetic Materials

The $M = \frac{dm}{dV}$ magnetization is defined as the magnetic dipole moment per unit volume.

The total current density is due to free and to magnetization current densities.

$$j = j_f + j_{m'}\ k = k_f + k_{m'}$$

(SI units) Gaussian units)

$$j_m = \vec{\nabla} \times M,\ k_m = M \times \hat{n}, \qquad j_m = c\vec{\nabla} \times M,\ k_m = cM \times \hat{n},$$

$$H = \frac{B}{\mu_0} - M \qquad H = B - 4\pi M$$

This definition is not unique.

For linear, isotropic, homogeneous magnetic materials we have:

(SI units)	(Gaussian units)
$M = \chi_m H$	$M = \chi_m H$
B = m0(H + M) = m0(1+cm)	B = H + 4pM = (1 + 4pcm)
H = m0 km H = mH	H = mH
$\vec{\nabla}\times H = j_f$	$\vec{\nabla}\times H = \frac{4\pi}{c}J_f$
$\chi_m < 0$, for diamagnetic materials,	$\chi_m < 0$, for paramagnetic materials.

BOUNDARY CONDITIONS FOR

$(H_2 - H_1).\hat{t} = k_f.\hat{n}$ (SI units), $(H_2 - H_1).\hat{t} = \frac{4\pi}{c}k_f.ft$ (Gaussian units),

$\vec{\nabla}.H \neq 0$ in general.

Maxwell's Equations

Maxwell's equations are:

(SI units)	(Gaussian units)
$\vec{\nabla}.E = \frac{\rho}{\varepsilon_0}, \vec{\nabla}\times E = -\frac{\partial B}{\partial t}$	$\vec{\nabla}.E = 4\pi\rho, \vec{\nabla}\times E = -\frac{1}{c}\frac{\partial B}{\partial t}$
$\vec{\nabla}.B = 0, \vec{\nabla}\times B = \mu_0 J + \frac{1}{c^2}\frac{\partial E}{\partial t}$	$\vec{\nabla}.B = 0, \vec{\nabla}\times B = \frac{4\pi}{c}J + \frac{1}{c}\frac{\partial E}{\partial t}$

or, in macroscopic form:

$\vec{\nabla}.D = \rho_f, \vec{\nabla}\times E = \frac{\partial B}{\partial t}$	$\vec{\nabla}.D = 4\pi\rho_f, \vec{\nabla}\times E = -\frac{1}{C}\frac{\partial E}{\partial t}$
$\vec{\nabla}.B = 0, \vec{\nabla}\times H = j_f + \frac{\partial D}{\partial t}$	$\vec{\nabla}.B = 0, \vec{\nabla}\times H = \frac{4\pi}{C}J_f + \frac{1}{c}\frac{\partial D}{\partial t}$

In lih materials with
$D = \varepsilon E$ and $B = \mu H$ and ε, μ = constant
we have:

$\vec{\nabla}.E = \frac{\rho f}{\varepsilon}, \vec{\nabla}\times E = \frac{\partial B}{\partial t}$	$\vec{\nabla}.E = \frac{4\pi}{\varepsilon}\rho f, \vec{\nabla}\times E = -\frac{1}{c}\frac{\partial B}{\partial t}$
$\vec{\nabla}.B = 0, \vec{\nabla}\times B = \mu j_f + \mu jf + \mu\varepsilon\frac{\partial E}{\partial t}$	$\vec{\nabla}.B = 0, \vec{\nabla}\times B = \frac{4\pi}{c}\mu j_f + \frac{\mu\varepsilon}{c}\frac{\partial E}{\partial t}$

Maxwell's equations are linear equations and the principle of superposition holds.

$$\vec{\nabla} . B = 0 \Rightarrow B = \vec{\nabla} \times A,$$

(SI units) (Gaussian units)

$$\vec{\nabla}\times\left(E+\frac{\partial A}{\partial t}\right)=0 \Rightarrow E+\frac{\partial A}{\partial t}=-\vec{\nabla}\phi \quad \vec{\nabla}\times\left(E+\frac{1}{c}\frac{\partial A}{\partial t}\right)=0 \Rightarrow E+\frac{1}{c}\frac{\partial A}{\partial t}=-\vec{\nabla}\phi$$

Faraday's Law

(SI units) Gaussian units)

$$\vec{\nabla}\times E = -\frac{\partial B}{\partial t} \qquad \vec{\nabla}\times E = -\frac{1}{c}\frac{\partial B}{\partial t}$$

$$\vec{\nabla}\times E.dr = -\frac{\partial}{\partial t}\int_A B.\hat{n}dA \qquad \vec{\nabla}\times E.dr = -\frac{1}{c}\frac{\partial}{\partial t}\int_A B.\hat{n}dA$$

Define the flux

$$F \equiv \int_A B.\hat{n}A$$

and the electromotive force $\varepsilon = \oint_\Gamma E.dr$.

Then:

$$\varepsilon = -\frac{\partial F}{\partial t} \qquad \varepsilon = -\frac{1}{C}\frac{\partial F}{\partial t}$$

Any induced emf tries to oppose the flux changes that produces it. This is Lenz's rule.

Quasi-static Situations

Consider N filamentary circuits. Then the flux through the ith circuit is $F_i = \sum_{j-1}^{N} F_{ij}$, where $F_{ij} = M_{ij}I_j$ (SI units), $F_{ij} = cM_{ij}I_j$ (Gaussian units). $M_{ij} =$ M_{ji} is the coefficient of mutual induction and $M_{ij} = L_i$ is the coefficient of self inductance. We have

$$\varepsilon_i = -\sum_j M_{ij}\frac{\partial I_j}{\partial t}$$

For a single filamentary circuit we have $\varepsilon_i = -L_i\frac{\partial I_j}{\partial t}$. To change the current in a circuit we need an external emf, V_{ext}, to overcome the induced emf □. $V_{ext} = -L\dfrac{\partial I}{\partial t}$

The energy stored in the circuit is $U = \dfrac{1}{2}LI^2$. For a system of N circuits we have:

(SI units) (Gaussian units)

$$U = \frac{1}{2}\sum_{m=1}^{N} F_m I_m, \qquad U = \frac{1}{2c}\sum_{m=1}^{N} F_m I_m, \text{ or}$$

$$U = \frac{1}{2c}\int j.A dV \qquad U = \frac{1}{2c}\int j.A dV \text{ or}$$

$$U = \frac{1}{2\mu_0}\int_{all\_space} B.B dV, \qquad U = \frac{1}{8\pi}\int_{all\_space} B.B dV$$

Energy and Momentum in Electrodynamics

Poynting's $E.j = -\dfrac{\partial u}{\partial t} - \nabla.S$ theorem, is a statement of energy conservation.

$u = \dfrac{1}{2\mu_0}B^2 + \dfrac{\varepsilon_0}{2}E^2$ (SI units), $u = \dfrac{1}{8\pi}(B^2 + B^2)$ (Gaussian units) is the energy density and

$S = \dfrac{1}{\mu_0}E \times B$ (SI units), $S = \dfrac{c}{4\pi}E \times B$ (Gaussian units) is the energy flux in the electromagnetic field. We define the momentum density as

$g = \dfrac{S}{c^2}$ (in SI and Gaussian units).

The Lorentz Gauge

If in electrodynamics we choose the Lorentz gauge defined through

(SI units) (Gaussian units)

$$\nabla.A = -\frac{1}{c^2}\frac{\partial \phi}{\partial t} \qquad \nabla.A = -\frac{1}{c}\frac{\partial \phi}{\partial t}$$

then□ and each Cartesian component of A satisfy the inhomogeneous wave equation

$$\left(\nabla^2 - \frac{1}{c^2}\frac{\partial^2}{\partial t^2}\right)\phi = -\frac{\rho}{\varepsilon_0} \qquad \left(\nabla^2 - \frac{1}{c^2}\frac{\partial^2}{\partial t^2}\right)\phi = -4\pi\rho$$

$$\left(\nabla^2-\frac{1}{c^2}\frac{\partial^2}{\partial t^2}\right)A=-\mu_0 j \qquad \left(\nabla^2-\frac{1}{c^2}\frac{\partial^2}{\partial t^2}\right)A=-\frac{4\pi}{c}j$$

$$\phi(r,t)=\frac{1}{4\pi\varepsilon_0}\int\frac{[\rho(r',t')]_{ret}\,dr'}{|r-r'|} \qquad \phi(r,t)=\int\frac{[\rho(r',t')]_{ret}\,dr'}{|r-r'|}$$

$$A_i(r,t)=\frac{\mu_0}{4\pi}\int\frac{[j_i(r',t')]_{ret}\,dr'}{|r-r'|} \qquad A_i(r,t)=\frac{1}{c}\int\frac{[j_i(r',t')]_{ret}\,dr'}{|r-r'|}$$

$$[\rho(r',t')]_{ret}=\rho\left(r',t-\frac{|r-r'|}{c}\right)$$

Choosing $\nabla.A=0$ is called choosing the Coulomb gauge.

The Wave Equation

In regions with $I_f=0$ *and* $j_f=0$ both E and B satisfy the wave equation.

(SI units) (Gaussian units)

$$\left(\nabla^2-\mu\varepsilon\frac{\partial^2}{\partial t^2}-\mu\sigma_c\frac{\partial}{\partial t}\right)E=0, \qquad \left(\nabla^2-\frac{\mu\varepsilon}{c^2}\frac{\partial^2}{\partial t^2}\right)E=0,$$

$$\left(\nabla^2-\mu\varepsilon\frac{\partial^2}{\partial t^2}\right)B=0 \qquad \left(\nabla^2-\frac{\mu\varepsilon}{c^2}\frac{\partial^2}{\partial t^2}\right)B=0.$$

Sinusoidal plane wave solutions $E=E(k.r-\omega t)$, $B=B(k.r-\omega t)$ exist.

We have $E\perp B, E\perp k, B\perp k, B=\sqrt{\mu\varepsilon}\hat{k}\times E$,

$k^2=\mu\varepsilon\omega^2$ (SI units), $k^2=\frac{\mu\varepsilon}{c^2}\omega^2$ (Gaussian units).

In regions with $I_f=0$ and $j_f=\tau_c E$ both E and B satisfy the damped wave equation.

(SI units) (Gaussian units)

$$\left(\nabla^2-\mu\varepsilon\frac{\partial^2}{\partial^2}-\mu\sigma_c\frac{\partial}{\partial t}\right)E=0, \qquad \left(\nabla^2-\frac{\mu\varepsilon}{c^2}\frac{\partial^2}{\partial^2}-\frac{4\pi}{c^2}\mu\sigma_c\frac{\partial}{\partial t}\right)E=0,$$

$$\left(\nabla^2-\mu\varepsilon\frac{\partial^2}{\partial^2}-\mu\sigma_c\frac{\partial}{\partial t}\right)E=0, \qquad \left(\nabla^2-\frac{\mu\varepsilon}{c^2}\frac{\partial^2}{\partial^2}-\frac{4\pi}{c^2}\mu\sigma_c\frac{\partial}{\partial t}\right)B=0,$$

Now

$$k^2=\iota\mu\sigma_c\omega+\mu\varepsilon\omega^2=\mu\varepsilon(\omega)\omega^2 \quad \left(k^2=i\frac{4\pi}{c^2}\mu\sigma_c\omega+\frac{\mu\varepsilon^2}{c^2}\omega^2-\frac{\mu\varepsilon(\omega)}{c^2}\omega^2\right)$$

i.e. k is complex, the wave damps out. The skin depth is defined as the distance it takes to reduce the amplitude by a factor of $\frac{1}{e}$.

Plane Waves at Boundaries

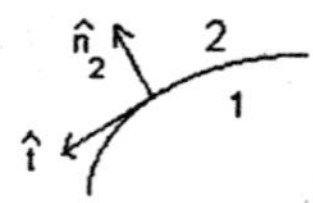

(SI units)	(Gaussian units)
$(D_2 - D_1).\hat{n}_2 = \sigma_f$	$(D_2 - D_1).\hat{n}_2 = 4\pi\sigma_f$
$(B_2 - B_1).\hat{n}_2 = 0,$	$(B_2 - B_1).\hat{n}_2 = 0,$
$(E_2 - E_1).\hat{t} = 0$	$(E_2 - E_1).\hat{t} = 0$
$(H_2 - H_1).\hat{t} = k_f.\hat{n}$	$(H_2 - H_1).\hat{t} = \frac{4\pi}{c} k_f.\hat{n}$

are the boundary conditions the electromagnetic fields have to satisfy on any boundary. For waves propagating across a dielectric-dielectric boundary or from a dielectric into a conductor we use $E.\hat{t}$ is continuous, $H.\hat{t}$ is continuous. From these boundary conditions we derive the law or reflection, and the law of refraction, (Snell's law), $n_1 \sin_i n_2 \sin_t$, for a wave propagating from a medium with index of refraction n_1 into a medium with index of refraction n_2. Here $\square_i$ $\square_r$ and $\square_t$ are the angles the incident, reflected, and transmitted wave vectors make with the normal to the boundary.

The plane of incidence is the plane containing the wave vector k_i and the normal to the boundary. A linear polarized wave has p-polarization if E lies in the plane of incidence, and it has s-polarization if E is perpendicular to the plane of incidence. The reflection coefficients for s and p polarization are

$$r_{12s} = \frac{1-\alpha,\beta}{1+\alpha,\beta}, \quad r_{12p} = \frac{\beta-\alpha}{\beta+\alpha},$$

and the transmission coefficients for s and p polarization are

$$t_{12s} = \frac{2}{1+\alpha\beta}, \quad t_{12p} = \frac{2}{\beta+\alpha}..$$

Here

$$\beta = \frac{n_2\mu_1}{n_1\mu_2} \text{ and } \alpha = \frac{\cos\theta_1}{\cos\theta_i}.$$

The reflectance is $R = \frac{<S_r>.\hat{n}}{<S_i>.\hat{n}} = |r_{12}|^2$, and the transmittance is

$$T = \frac{<S_r>.\hat{n}}{<S_i>.\hat{n}} = \alpha,\beta\,|r_{12}|^2.$$

We have $R+T=1$ (energy conservation).
For conductors n_2, ε_2, and $\cos\theta_t$ are complex numbers.

Polarization by Reflection

When $\theta_i = \theta_B$, the Brewster angle, then only light with s-polarization is reflected. We have

$$\tan\theta_B = \frac{n_2}{n_1},\quad \theta_B + \theta_i = \frac{\pi}{2}.$$

Total Internal Reflection

If $n_2 < n_1$, then for

$$\theta_i > \theta_c = \frac{n_2}{n_1}$$

we have total internal reflection. The wave vector k_t then has a component parallel to the interface which is real and a component perpendicular to the interface which is imaginary.

Reflection by a Good Conductor

For good conductors $R \approx 1$, $R \approx 0$. The wave does not penetrate into the conductor. We then use the boundary $E.\hat{t}$ conditions is continuous, $B.\hat{n}$ is continuous.

Electrodynamics in Relativistic Notation

A contravariant 4-vector is a set of 4 quantities which transform under a Lorentz transformation like (ct, r) = (x^0, x^1, x^2, x^3). (A^0, A^1, A^2, A^3) is a contravariant 4-vector if

$$A'^{\alpha} = \frac{\partial X'^{\alpha}}{\partial X^{\beta}} A^{\beta}.$$

A covariant 4-vector is a set of 4 quantities which transform under a Lorentz transformation like (ct, $-r$) = (x_0, x_1, x_2, x_3). (A_0, A_1, A_2, A_3) is a

contravariant 4-vector if $A'_{\alpha} = \frac{\partial X^{\beta}}{\partial X'^{\alpha}} A_B$.

If the primed coordinate system moves with velocity with respect to the unprimed one, then

$$\begin{pmatrix} x^{0'} \\ x^{1'} \\ x^{2'} \\ x^{3'} \end{pmatrix} = \begin{pmatrix} y & -y\beta & 0 & 0 \\ -y\beta & y & 0 & 0 \\ 0 & 0 & 1 & 0 \\ 0 & 0 & 0 & 0 \end{pmatrix} \begin{pmatrix} x^{0} \\ x^{1} \\ x^{2} \\ x^{3} \end{pmatrix}, \quad \begin{pmatrix} x_0' \\ x_1' \\ x_2' \\ x_3' \end{pmatrix} = \begin{pmatrix} y & y\beta & 0 & 0 \\ y\beta & y & 0 & 0 \\ 0 & 0 & 1 & 0 \\ 0 & 0 & 0 & 0 \end{pmatrix} \begin{pmatrix} x_0 \\ x_1 \\ x_2 \\ x_3 \end{pmatrix}.$$

$$\beta = \frac{\mathrm{v}}{\mathrm{C}},\ \beta = \frac{\mathrm{v}}{\mathrm{C}},\ y = \left(1-\beta^2\right)^{-\frac{1}{2}}.$$

A contravariant tensor of second rank is a set of 16 quantities which transform under a Lorentz transformation according to

$$F'^{\alpha\beta} = \frac{\partial x'^{\alpha}}{\partial x^{y}} \frac{\partial x'^{\beta}}{\partial x^{\delta}} F^{y\delta}.$$

A covariant tensor of second rank transforms under a Lorentz transformation according to

$$G'_{\alpha\beta} = \frac{\partial x^{y}}{\partial x'^{a}} \frac{\partial x^{\delta}}{\partial x'^{\beta}} G_{y\delta}$$

and a mixed tensor transforms according to

$$H'^{\alpha}_{\beta} = \frac{\partial x'^{\alpha}}{\partial x^{y}} \frac{\partial x^{\delta}}{\partial x'^{\beta}} H^{y}_{\delta}.$$

Special tensors:

$$\delta'^{\alpha}_{\beta} = \frac{dx^{\alpha}}{dx^{\beta}}$$

is the Kroneker delta extended to 4 indices.

$$g^{\alpha\beta} = g_{\alpha\beta} = \begin{pmatrix} 1 & 0 & 0 & 0 \\ 0 & -1 & 0 & 0 \\ 0 & 0 & -1 & 0 \\ 0 & 0 & 0 & -1 \end{pmatrix}$$

is the metric tensor.

The dot product between two contravariant 4-vectors is defined as $A.B = g_{\alpha\beta}A^{\alpha}B^{\beta} = A_{\beta}B^{\beta} = A^{\alpha}B_{\alpha}$. It is invariant under a Lorentz transformation, it is a Lorentz scalar.

Important 4-vectors:

$$X^{\mu} = (ct, r), \quad \underbrace{u^{\mu} = (yc, yv)}_{\text{4-vector velocity}} \quad \underbrace{p^{\mu} = (ymc, ymv) = \left(\tfrac{E}{c}, p\right) = (p^{0}, p),}_{\text{4-vector momentum}}$$

$$\frac{\partial}{\partial x_\mu} = \partial^u = \left(\frac{\partial}{\partial x_0}, -r\right) \text{ 4-dimensional gradient}$$

(SI units),

$$\underbrace{j^\mu = (c\rho, j)}_{\text{4-vector current}} \qquad \underbrace{A^\mu = \left(\tfrac{\phi}{c}, A\right)}_{\text{4-vector potential}}$$

$j^\mu = (c\rho, j)$, $A^\mu = (\phi, A)$ (Gaussian units).

The divergence of a 4-vector

$$\partial^\mu A_\mu = \partial_\mu A^\mu = \frac{\partial A^0}{\partial x^0} - \bar{\nabla}.A$$

is a Lorentz scalar.

Examples: $\partial^\mu j_\mu = \partial_\mu j^\mu = \frac{1}{c^2}\frac{\partial(c\rho)}{\partial(ct)} - \bar{\nabla}.j = 0$ is the statement of charge conservation.

$$\partial^\mu A_\mu = \partial_\mu A^\mu = \frac{1}{c^2}\frac{\partial\phi}{\partial t} + \bar{\nabla}.A = 0 \text{ (SI)},$$

$\partial^\mu A_\mu = \partial_\mu A^\mu = \frac{1}{c}\frac{\partial\phi}{\partial t} + \bar{\nabla}.A = 0$ (Gaussian), is the Lorentz condition.

All contravariant 4-vectors transform as

$$A^0, = y(A^0 - \beta.A), A_{\|}' = y(A_{\|} - \beta A^0), A_{\perp}' = A_{\perp}.$$

The antisymmetric field strength tensor is defined through

$$F^{\alpha\beta} = \partial^\alpha A^\beta - \partial^\beta A^\alpha.$$

It is a second rank contravariant tensor.

$$F^{\alpha\beta} = \begin{pmatrix} 0 & \frac{-E_x}{c} & \frac{-E_y}{c} & \frac{-E_z}{c} \\ \frac{E_x}{c} & 0 & -B_z & B_y \\ \frac{E_y}{c} & B_z & 0 & -B_x \\ \frac{E_c}{c} & -B_y & B_x & 0 \end{pmatrix} \text{ (SI units)},$$

$$F^{\alpha\beta} = \begin{pmatrix} 0 & -E_x & -E_y & -E_z \\ E_x & 0 & -B_z & B_y \\ E_y & B_z & 0 & -B_x \\ E_z & -B_y & B_x & 0 \end{pmatrix} \text{ (Gaussian)},$$

$F'^{\alpha\beta} = \dfrac{\partial x'^{\alpha}}{\partial x^{y}} \dfrac{\partial x'^{\beta}}{\partial x^{\delta}} F^{y\delta}$. This yields

$E'_{\parallel} = E_{\parallel},\ E'_{\parallel} = E_{\parallel}$ (SI units),

$E'_{\parallel} = E_{\parallel},\ E'_{\parallel} = E_{\parallel}$ (Gaussian units),

$E'_{\perp} = y(B + v \times B)_{\perp}, B'_{\perp} = y\left(B - \dfrac{v}{c^2} \times E\right)_{\perp}$ (SI units),

$E'_{\perp} = y\left(E + \dfrac{v}{c} \times B\right)_{\perp}, B'_{\perp} = y\left(B - \dfrac{v}{c} \times E\right)_{\perp}$ (Gaussian units).

For a charge distribution moving with velocity v with respect to an observer, we have in the frame K' of the observer

$E'_{\parallel} = E'_{\parallel},\ B'_{\parallel} = 0,$ (SI units),

$E'_{\parallel} = E_{\parallel},\ B'_{\parallel} = 0,$ (Gaussian units),

$E'_{\perp} = yE_{\perp},\ B'_{\perp} = y\left(\dfrac{v}{c^2} \times E\right)_{\perp},$ (SI units),

$E'_{\perp} = yE_{\perp},\ B'_{\perp} = y\left(\dfrac{v}{c} \times E\right)_{\perp},$ (Gaussian units),

or $B'_{\perp} = y\dfrac{v}{C^2} \times E_{\perp} = \dfrac{v}{C^2} \times E'_{\perp},$ (SI units)

$B'_{\perp} = y\dfrac{v}{C} \times E_{\perp} = \dfrac{v}{C^2} \times E'_{\perp},$ (Gaussian units),

or $B' = \dfrac{v}{C} \times E'.$ (SI units),

$E^2 - c^2 B^2$ (Gaussian units).

$E^2 - c^2B^2$ (SI units) or $E^2 - B^2$ (Gaussian units) and $(E \cdot B)^2$ are invariant under a Lorentz transformation.

Radiation Produced by Moving Charges

The radiation field $E(r, t)$ of a point charge moving *nonrelativistically* is

$$E(r,t) = -\left[\frac{1}{4\pi\varepsilon_0}\right]_{SI} \frac{q}{c^2 r''} a_{\perp}\left(t - \frac{r''}{c}\right),$$

where

$$r'' = r - r'\left(t - \frac{|r - r'|}{c}\right),$$

i.e. the vector from the charge to the observer at the retarded time $t - \frac{|r - r'|}{c}$.

$$B = \frac{\hat{r}''}{c} \times E \text{ (SI units)}, \quad B = \hat{r}'' \times E \text{ (Gaussian units)}.$$

For the near field it is a better approximation to use the instantaneous Coulomb field than to use the retarded Coulomb field.

The energy flux associated with the fields of a point charge is calculated from the Poynting vector S. The total power radiated is

$$P = \oint S.\hat{n}dA = \frac{2}{3}\frac{e^2 a^2}{c^3}$$

in SI and Gaussian units, with $e^2 = \frac{q^2}{4\pi\varepsilon_0}$ in SI units. This is the Lamor formula.

The Liénard-Wiechert potentials

The potentials of a point charge moving in an arbitrary way are
(SI units) (Gaussian Units)

$$\phi(r,t) = \frac{1}{4\pi\varepsilon_0}\left[\frac{q}{(1-\beta.\hat{n})R}\right]|ret, \quad \phi(r,t) = \left[\frac{q}{(1-\beta.\hat{n})R}\right]|ret,$$

$$A(r,t) = \frac{1}{4\pi\varepsilon_0 c}\left[\frac{q\beta}{(1-\beta.\hat{n})R}\right]|ret, \quad A(r,t) = \left[\frac{q\beta}{(1-\beta.\hat{n})R}\right]|ret.$$

Here $\hat{n}$R is the vector pointing from the point charge to the observer at r. The potentials of a point charge depend only on the position and the velocity at the retarded time. The fields E and B depend on the acceleration.

The Equations of Motion

Neglecting radiation, the equations of motion of a particle of charge q in external fields E and B can be written as
(SI units) (Gaussian units)

$$\frac{dp}{dt} = q(E + v \times B), \quad \frac{dE}{dt} = qv.E, \qquad \frac{dp}{dt} = q\left(E + \frac{v}{c} \times B\right), \quad \frac{dE}{dt} = qv.E,$$

or in covariant for

$$\frac{du^{\mu}}{d\tau} = \frac{q}{m} F^{\mu\upsilon} u_{\upsilon}, \qquad \frac{du^{\mu}}{d\tau} = \frac{q}{m} F^{\mu\upsilon} u_{\upsilon},$$

Here $u^{\mu} = (yC, yv) = \frac{p^{u}}{m}$.

Radiation Produced by Moving Charges

The radiation field $E(r,t)$ of a point charge moving *nonrelativistically* is

$$E(r,t) = -\left[\frac{1}{4\pi\varepsilon_0}\right]_{SI} \frac{q}{C^2 r''} a_{\perp}\left(t - \frac{r''}{C}\right),$$

where

$$r'' = r - r'\left(t - \frac{|r - r'|}{c}\right),$$

i.e. the vector from the charge to the observer at the retarded time $t - \frac{|r - r'|}{c}$.

$$B = \frac{\hat{r}''}{c} \times E \text{ (SI units)}, B = \hat{r}'' \times E \text{ (Gaussian units)}.$$

For the near field it is a better approximation to use the instantaneous Coulomb field than to use the retarded Coulomb field.

The energy flux associated with the fields of a point charge is calculated from the Poynting vector S. The total power radiated is

$$P = \oint S.\hat{n} dA = \frac{2}{3}\frac{e^2 a^2}{c^3}$$

in SI and Gaussian units, with $e^2 = \frac{q^2}{4\pi\varepsilon_0}$ in SI units. This is the Lamor formula.

The Liénard-Wiechert Potentials

The potentials of a point charge moving in an arbitrary way are (SI units) (Gaussian Units)

$$\phi(r,t) = \frac{1}{4\pi\varepsilon_0}\left[\frac{q}{(1 - \beta.\hat{n})R}\right] | ret, \qquad \phi(r,t) = \left[\frac{q}{(1 - \beta.\hat{n})R}\right] | ret,$$

$$A(r,t) = \frac{1}{4\pi\varepsilon_0 c}\left[\frac{q\beta}{(1-\beta.\hat{n})R}\right]\Big|ret, \quad A(r,t) = \left[\frac{q\beta}{(1-\beta.\hat{n})R}\right]\Big|ret,$$

Here $R\hat{n}$ is the vector pointing from the point charge to the observer at r. The potentials of a point charge depend only on the position and the velocity at the retarded time. The fields E and B depend on the acceleration.

The equations of motion: Neglecting radiation, the equations of motion of a particle of charge q in external fields E and B can be written as

(SI units) (Gaussian units)

$$\frac{dp}{dt} = q(E + v \times B), \quad \frac{dE}{dt} = qv.E, \qquad \frac{dp}{dt} = q\left(E + \frac{v}{C} \times B\right), \quad \frac{dE}{dt} = qv.E,$$

or in covariant for

$$\frac{du^{\mu}}{d\tau} = \frac{q}{m}F^{\mu\upsilon}u_{\upsilon}, \qquad \frac{du^{\mu}}{d\tau} = \frac{q}{m}F^{\mu\upsilon}u_{\upsilon},$$

Here $u^{\mu} = (yc, yv) = \frac{p^{u}}{m}$.

AC circuits

Assume $V(t)$, $I(t)$, (t) are all proportional to e^{taxt}.

Assume idealized circuit elements. Define the impedance Z as $Z = \frac{V}{I}$.

Then Z (inductance) = $Z_L = i\omega L$,

Z (capacitance) = ZC = $\frac{1}{i\omega C}$,

Z (resistance) = $Z_R = R$.

Kirchhoff's rules for filamentary circuits:

For each loop $\sum_n V_n = 0$, for each node $\sum_n I_n = 0$.

Thévenin equivalent circuits: Any two terminal network can be replaced by a generator? I_{eff} in series with an impedance Z_{eff}.

Norton equivalent circuits: Any two terminal network can be replaced by a current source I_{eff} in parallel with an impedance Z_{eff}.

The maximum power is delivered to a load when $Z_{load} = Z_{eff}$.

Ladder networks: Consider an infinite ladder network of elements as shown.

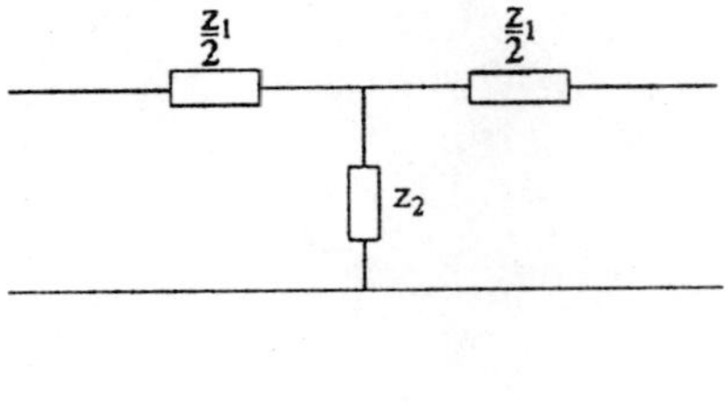

The network is equivalent to a network terminated by Z_0, if $Z_0 = \sqrt{\frac{Z_1^2}{4} + Z_1 Z_2}$.

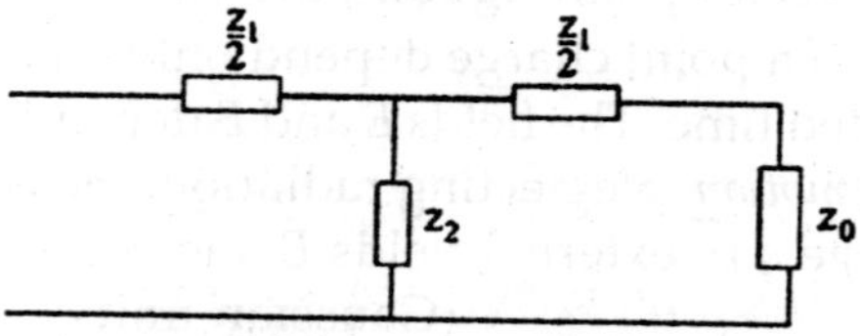

If Z_0 is real the circuit absorbs energy, if Z_0 is imaginary, it does not absorb energy. If Z_0 is real signals can pass to the load, if Z_0 is imaginary, signals cannot pass to the load, we have a filter.

Chapter 5

Ampere's Law, Solenoid, Magnetic Field, Diamagnetic

MAGNETIC FIELDS PRODUCED BY CURRENTS

1. A long straight wire carrying a current:
   Can a current carrying wire produce a magnetic field?
   Answer: Yes. Has been observed experimentally
   *Hans Christian Oersted first discovered the effect in 1830. *
   Figure (a) illustrates Oersted's discovery with a very long wire.

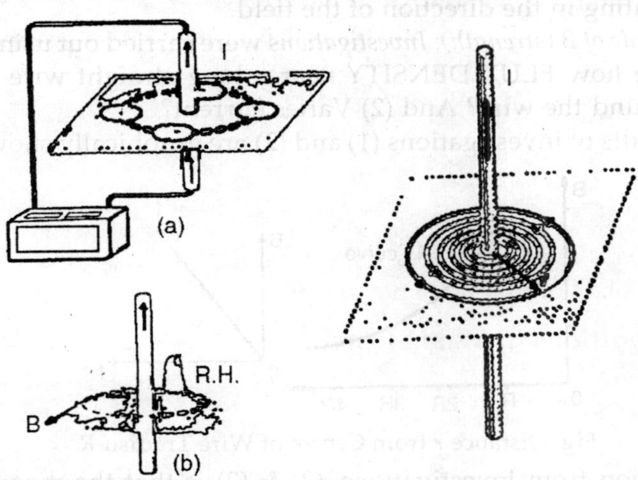

Observations: A compass needle placed in the vicinity of the current carrying wire deflects in a certain manner. Compass needle points in a circular pattern about the wire.

The circular pattern disappears when the current drops to zero.

Compass needles interact with magnetic fields. So the observations lead to the conclusion that a current carrying wire can produce a magnetic field.

Another observation is the fact that if the direction of the current is reversed, the needle reverses its pointing direction indicating that the direction of the field has reversed.

This leads to the conclusion that the direction of the magnetic field is dependent on the direction of the current.

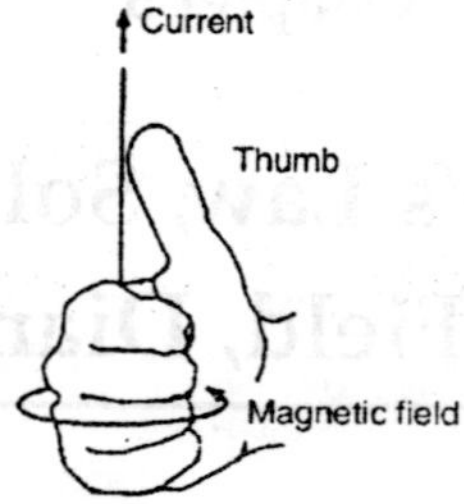

1. The direction of the field can be found by using the Cork Screw Rule Or The Right-Hand Grip Rule.
   Cork Screw Rule: Imagine driving a corkscrew into a cork. Its rotation is in the same direction as the field lines of the wire carrying current in the driving direction.
2. *Right-hand grip rule:* Grip the wire using the right hand with the thumb pointing in the direction of the current -- the fingers then pointing in the direction of the field.

Magnitude of B (strength): Investigations were carried out using the Hall Probe to see how FLUX DENSITY near a long straight wire (1) varies position around the wire? And (2) Varies current?

The results of investigations (1) and (2) are graphically shown below.

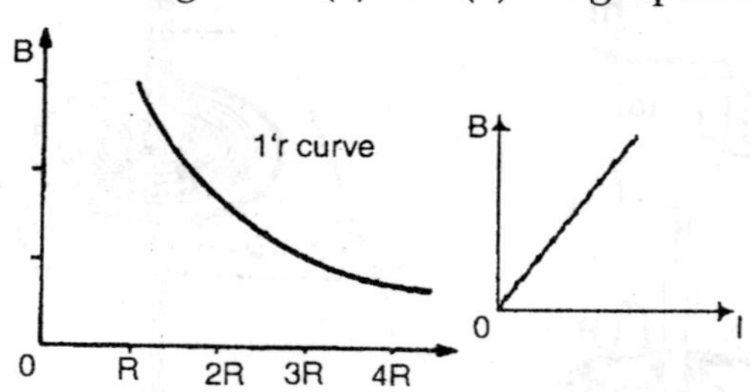

Fig. Distance r from Centre of Wire Tradisu R

Conclusion from investigations, (2) & (3) is that the magnetic filed around a long straight is proportional to the current flowing through it and inversely proportional to the perpendicular distance from the centre of the wire to the probe (or a point P). The final equation for the flux density:

$$B = \frac{\mu_0 I}{2\pi r}$$

Where B = the flux density (T)

I = the current in the wire (A)

r = the distance of P from the wire

m_0 = the PERMEABILITY of free space (vacuum) 4p × 10^{-7} T m A

Field lines are closer near the wire and the field is stronger, as number of lines per unit area is proportional to the strength.

Example 1: A particle with charge 6.5 × 10^{-6}C moves parallel to a wire carrying a current of 3.0 A. The distance of the particle from the wire is 0.05 m, and the speed of the particle is 280 m s^{-1}. Determine the magnetic force (magnitude and direction) experienced by the particle if the direction of motion of the particle is,

(*i*) Same as that of the current and

(*ii*) Opposite to that of the current.

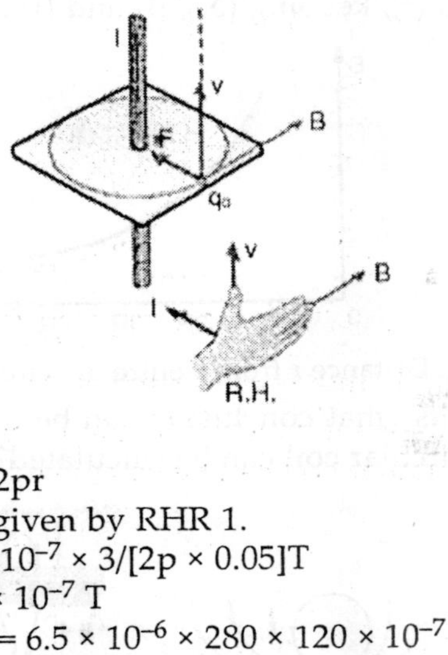

Force = q_ovB

Magnitude, B = m_oI/2pr

Direction of force given by RHR 1.

B = 4p × 10^{-7} × 3/[2p × 0.05]T

= 120 × 10^{-7} T

(*i*) v up: F = q_ovB = 6.5 × 10^{-6} × 280 × 120 × 10^{-7} N

= 2.18 × 10^{-8} N

Direction is inwards (towards the wire), as shown.

(*ii*) V is down: F = 2.18 × 10^{-8} N, Direction outwards (away from wire).

FLUX DENSITY AT THE CENTRE OF A FLAT CIRCULAR COIL

The field pattern due to a current in a flat circular coil is shown above. The field direction is given by using the RIGHT HAND GRIP RULE.

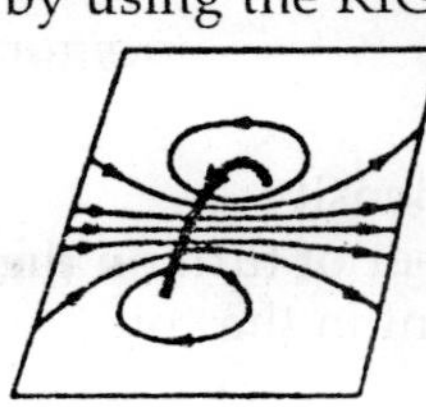

Investigations were carried out using the hall probe to see how the MAGNETUDE OF FLUX DENSITY (B) at the centre of the coil.

(1) Varies with current (I), when (N) the number of turns of the coil and (r) the radius of the coil are constant.

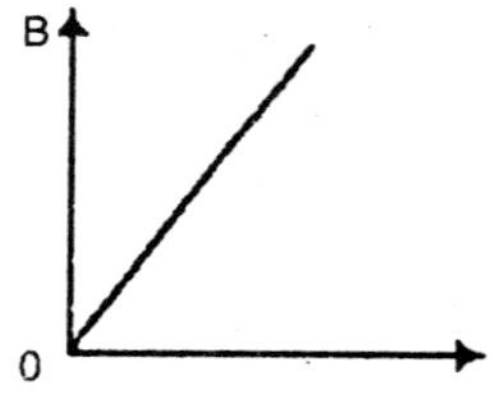

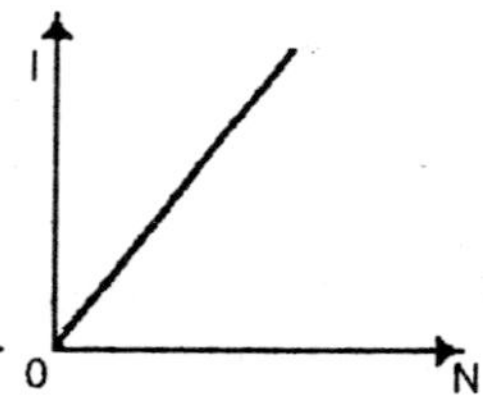

(2) Varies with (N), when (r), (*l*) the length of the wire making up the coil and (I) are constant.

(3) Varies with (r), keeping (N), (*l*) and (I) constant.

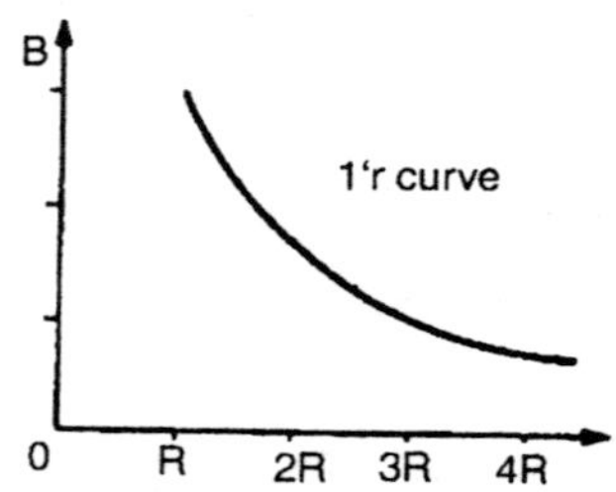

Fig. Distance *r* from Centre of Wire Tradius.

From the graphs what conclusion can be made about how the flux density (B) in flat circular coil can be calculated?

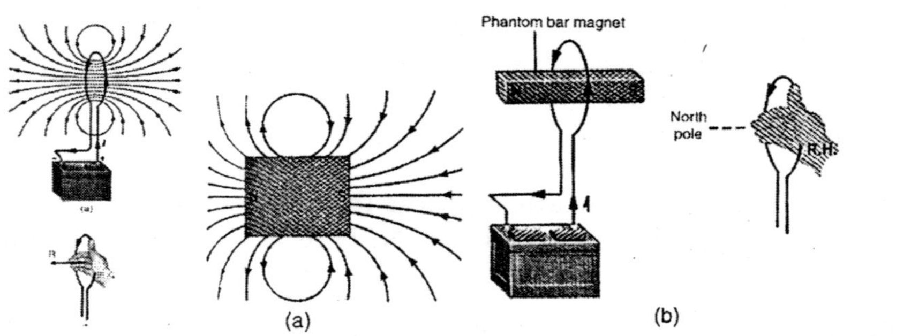

The magnitude of the flux density at the centre of flat circular coil is given by

$$B = \frac{\mu_0 NI}{2r}$$

Where B = the flux density

N = the number of turn on the coil

I = the current in the coil

r = the radius of the coil

μ_0 = the permeability of free space (= $4\pi \times 10^{-1}Hm^{-1}$)

2.Field due to a loop carrying a current

B at the centre = $m_0I/2R$, R is the radius of the loop.

For a flat coil of N loops, B at the centre = $\mu_0NI/2R$

FLUX DENSITY DUE TO A LONG SOLENOID

The magnetic field pattern due to a current in a long solenoid is shown above. The pattern is similar to that of a bar magnet, the field line spread from the North Pole, loop around, then converge on the South Pole. The direction of the field associated with the solenoid can be determined by using the Right Hand Grip Rule.

To determine the polarity of a solenoid, given the current direction, view the left end of the solenoid shown above the direction of the current is clockwise, that end is a south pole; viewing right end shows an anticlockwise current direction, which constitutes a north pole.

1. The variation of the flux density across and along the solenoid keeping the current (I) through the coil constant, the number of turns (N) of coil and (l) length constant.

   The result: The flux density is fairly constant near the centre of the solenoid and *half it centre value* at its ends.

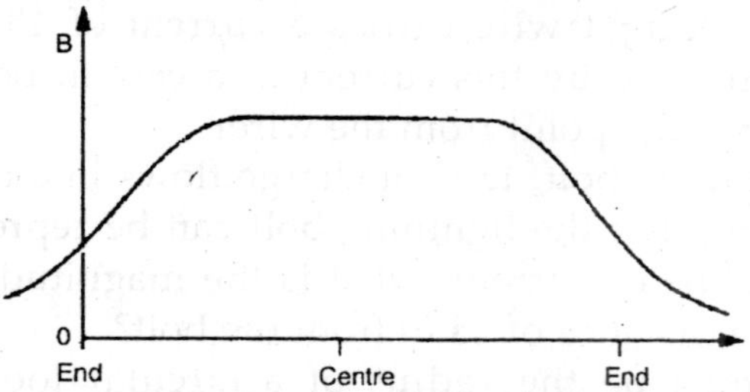

2. The variation of the flux density with current (I). When (N), (l) and (d) the solenoid diameter is constant.

   The result shown graphically that the flux density is proportional to the current.

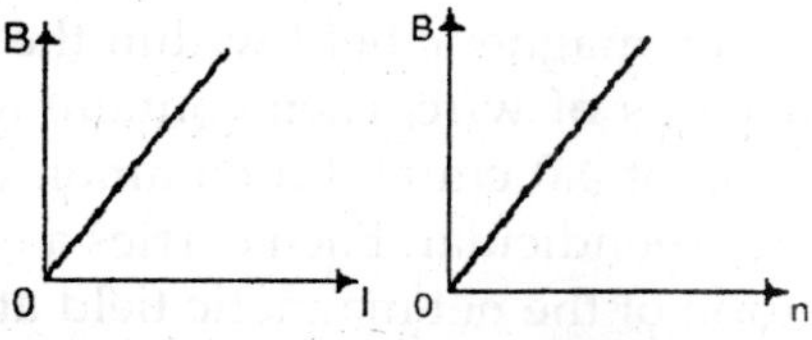

3. The variation of magnetic field strength with the number of turns per metre (n), keeping (I), (l) and (d) constant.

   The result: A graph of magnetic field strength against (n) per metre gives a straight line through the origin.

Conclusion: The flux density is proportional to (n) the number of turns per metre.

4. The variation of the flux density with the solenoid's diameter, keeping (I), (l) and (N) constant.

What would be the result of the investigation?

The variation of magnetic field strength with solenoid diameter: To measure this, you will need to use several metres of insulated wire to make solenoids of different diameters. Each solenoid has the same length and number of turns. Pass the same current through each solenoid, and measure the magnetic field strength inside. Your measurements ought to show that the diameter of the solenoid does not affect the magnetic field strength inside it.

The magnitude of the flux density near the centre of a long solenoid is given by

$B = \mu_0 nI$ B = the flux density (T)
n = the number of turns per metre (m–1)
I = the current through the solenoid (A)

(i) $B = \frac{\mu_0 NI}{L} - \mu_0 nI$

(ii) A either end the flux density is B = m_0 nI/2

1. A long, straight wire carries a current of 48 A. The magnetic field produced by this current at a certain point is 8.0 ´ 10^{-5} T How far is the point from the wire?
2. In a lightning bolt, 15 C of charge flows in a time of 1.5 ´ 10^{-3} s. Assuming that the lightning bolt can be represented as a long, straight line of current, what is the magnitude of the magnetic field at a distance of 25 m from the bolt?
3. What must be the radius of a circular loop of wire so the magnetic field at its centre is 1.8 ´ 10^{-4} T when the loop carries a current of 12 A?
4. A long solenoid consists of 1400 turns of wire and has a length of 0.65 m. There is a current of 4.7 A in the wire. What is the magnitude of the magnetic field within the solenoid?
5. Two circular loops of wire, each containing a single turn, have the same radius of 4.0 cm and a common centre. The planes of the loops are perpendicular. Each carries a current of 1.7 A. What is the magnitude of the net magnetic field at the common centre?
6. Two long, straight wires are separated by 0.120 m. The wires carry currents of 8.0 A in opposite directions, as the drawing indicates. Find the magnitude of the net magnetic field at the points labelled (a) A and (b) B.

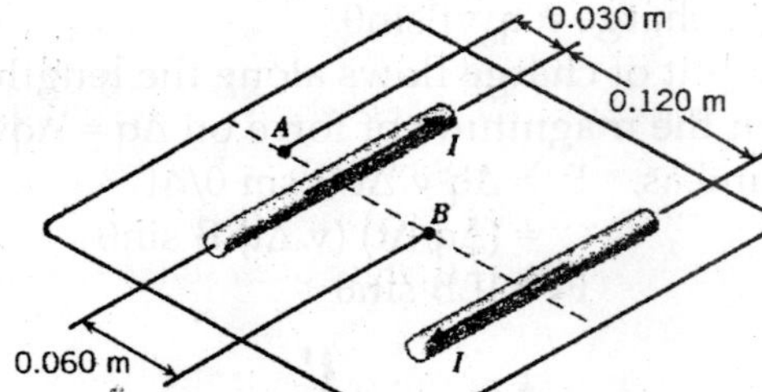

7. Two circular coils are concentric and lie in the same plane. The inner coil contains 120 turns of wire, has a radius of 0.012 m, and carries a current of 6.0 A. The outer coil contains 150 turns and has a radius of 0.017 m. What must be the magnitude and direction (relative to the current in the inner coil) of the current in the outer coil, such that the net magnetic field at the common centre of the two coils is zero?
8. A piece of copper wire has a resistance per unit length of 5.90 ′ 10^{-3} W/m. The wire is wound into a thin, flat coil of many turns that has a radius of 0.140 m. The ends of the wire are connected to a 12.0-V battery. Find the magnetic field strength at the centre of the coil.
9. The drawing shows an end-on view of three wires. They are long, straight, and perpendicular to the plane of the paper. Their cross sections lie at the corners of a square. The currents in wires 1 and 2 are $I_1 = I_2 = I$ and are directed into the paper. What is the direction of the current in wire 3, and what is the ratio I_3/I, such that the net magnetic field at the empty corner is zero?

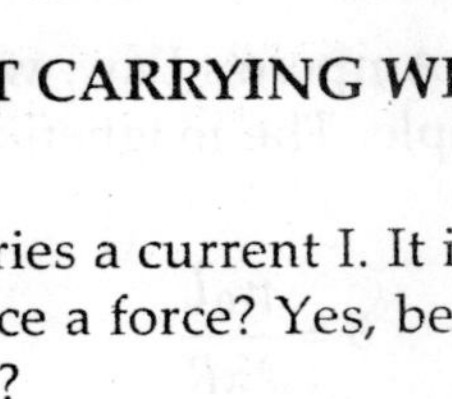

FORCE ON A CURRENT CARRYING WIRE (CONDUCTOR) IN A MAGNETIC FIELD

A wire of length L carries a current I. It is oriented at θ to B.

Will the wire experience a force? Yes, because current constitutes of moving charges. What is it?

The magnetic force exerted on the wire is the net force acting on the total amount of charge moving in the wire.

Force on a single charge = $q_o vB\sin\theta$.

Suppose Δq amount of charge flows along the length of the wire in a time interval Δt, then the magnitude of force on $\Delta q = \Delta qvB\sin\theta$.

Which is the same as, $F = \Delta q\ v\ \Delta t\ B\ \sin\theta/\Delta t$

$$= (\Delta q/\Delta t)\ (v\ \Delta t)\ B\ \sin\theta$$

$$F = ILB\ \sin\theta$$

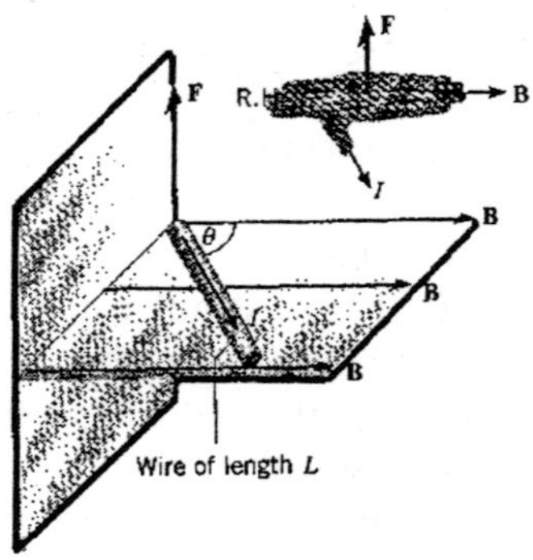

Force is maximum and is equal to ILB, when $\theta = 90°$ and 0 when $\theta = 0°, 180°$.

AMPERE'S LAW (SPECIAL CASE OF BIOT-SAVART LAW)

The Biot-Savart Law is very powerful. It will work for any current carrying wire regardless of the shape and it doesn't require you to know anything about the shape of the magnetic field lines. However, the math required to solve the Biot-Savart equation is difficult and often requires numerical techniques.

When we were solving electric field problems, we faced a similar problem and discovered that we could solve many electrostatic problems with far less math using Gauss' Law of Electric Fields as long as the problem exhibited symmetry (i.e. we knew the shape of the field).

In magnetic fields, we have a similar situation in which the magnetic field can be determined using far less difficult math as long as the system exhibits symmetry and we *know* the basic shape of the magnetic field *before* we do the problem.

To heuristically develop Ampere's Law, we consider our results for the infinitely long wire example. The magnetic field strength was given by

$$B = \frac{\mu_0 I}{2\pi R}$$

After rearranging the equation, we have

$$B\ (2\pi R) = \mu_0 I$$

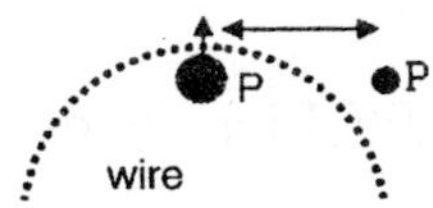

Fig. Wire

From calculus, we know that for the circular path we have

$$2\pi R = \oint R d\theta = \oint ds$$

Thus, we have

$$B\oint ds = \mu_0 I$$

and since B is constant and parallel to the path of integration (i.e. to the displacement vector,) we find that

$$\oint \vec{B} \bullet d\vec{s} = \mu_0 I_p$$

Where I_p is the total net current (in our case I) that penetrates the surface bounded by our integration curve (in our case a circle).

Ampere's Law is very similar to Gauss's Law in electrostatics is that it can greatly simplify our math work but can only be used on problems that exhibit a great amount of symmetry.

Ampere's Law is useful if the following conditions are met:

1. You know the basic shape of the B-filed before you do the problem
2. An integration curve can be found in which the B-field is either perpendicular to the path or is constant in magnitude and parallel to the integration path.

MAGNETIC DIPOLE REVISITED

In order to take advantage of Ampere's Law, we must be able to predict the shape of the magnetic field for our system (i.e. know its symmetry). The magnetic dipole (current loop) is the basic building block for several important engineering structures including the solenoid (coil or inductor) and the toroid.

Very Long Solenoid (Coil With Many Loops)

We are now ready to use Ampere's Law and our work with the current loop to solve for the magnetic field due to a very long coil (i.e. when the length of the coil is much greater than the coil's diameter). In the figure below, we have a coil with N turns (loops) and a current I. The cross sectional view of the coil is also shown.

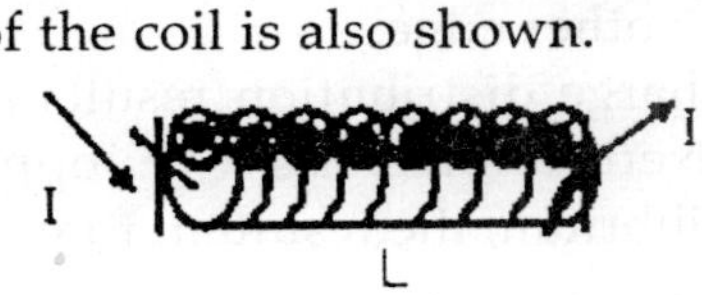

Field of a Long Solenoid

PURPOSE: To show the field inside, outside, and at the ends of a long solenoid.

DESCRIPTION: A long solenoid is supplied with a DC current and is probed with a Hall probe, as shown in the above left photo. Use an axial probe to investigate the field inside the solenoid, as shown below left, and show that it is uniform in magnitude and direction. Probe the field outside the solenoid, as shown below right, and show there is no B field.

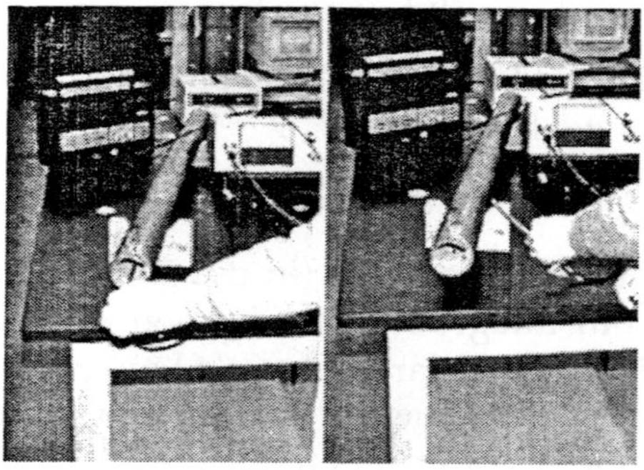

Another long solenoid is placed onto the overhead and powered by a power supply. A clear magnetic compass is used to probe the magnetic field at the end of the solenoid.

EQUIPMENT: The coil and magnetic compass are stored in the electricity and magnetism cabinet, the Hall Probe is borrowed from the 2nd floor labs. Be sure to get the axial probe and the F.W. Bell model 620 Gaussmeter.

SETUP NOTES: Use either the VOM meter or the electrometer to display the hall measurement to the class.

THE THEORY OF THE HALL EFFECT

A particle with charge Q, velocity, $\bar{V}$ and moving within a magnetic field, $\bar{B}$, will experience the Lorentz force, $F = Q(\bar{V} \times \bar{B})$. The force direction is mutually perpendicular to the directions of the particle velocity and the magnetic field. If a long, flat current-carrying conductor is placed in a magnetic field, the moving charges will experience a net force mutually perpendicular to the direction of the current flow (longitudinal conductor axis) and the magnetic field. Under the influence of this force, the electrons will "pile up" on one edge of the conductor, and positive charges will gather on the other edge.

An uneven lateral charge distribution results and gives rise to an electric field, $\bar{E}$, which exerts a force, $\bar{F} = Q\bar{E}$, opposite in direction to the Lorentz force. At equilibrium, the resultant forces balance. This field, superimposed on the E in the direction of the current flow, yields the skewed equipotential lines first noted by Hall. The relation between the voltage, current, and magnetic field can be generalized as follows:

VH = y IB
VH = Hall voltage
y = a constant product sensitivity
I = Hall current
B = magnetic field perpendicular to Hall Plate surface

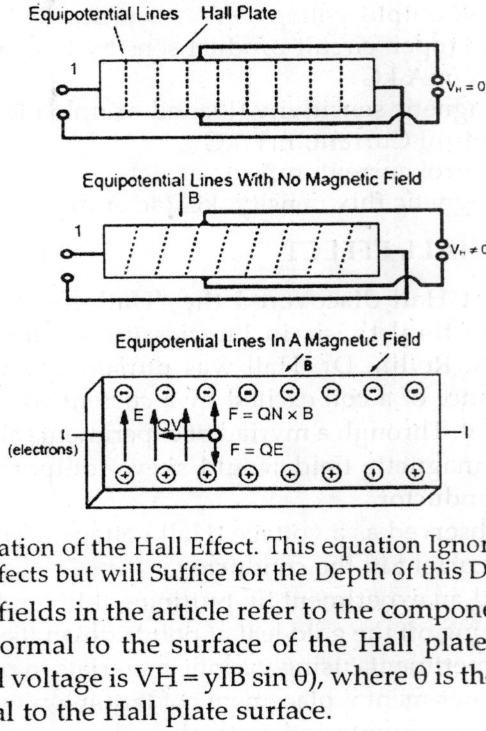

Fig. Explanation of the Hall Effect. This equation Ignores many low Level Effects but will Suffice for the Depth of this Discussion.

Note: All B fields in the article refer to the component of the external B field that is normal to the surface of the Hall plate. A more general equation for Hall voltage is VH = yIB sin θ), where θ is the angle is between B and the normal to the Hall plate surface.

HALL SENSOR

A Hall sensor is a four-terminal, solid-state device capable of producing an output voltage VH, proportional to the product of the input current, lc, the magnetic flux density, B, and the sine of the angle betweeen B and the plane of the Hall sensor. A reversal in the direction of either the magnetic field or the control current will result in a polarity change of VH. A reversal in the direction of both will keep the polarity the same. By holding the control current constant, the Hall voltage may be used to measure magnetic flux density. Multiplication may be accomplish by varying both the control current and the magnetic field.

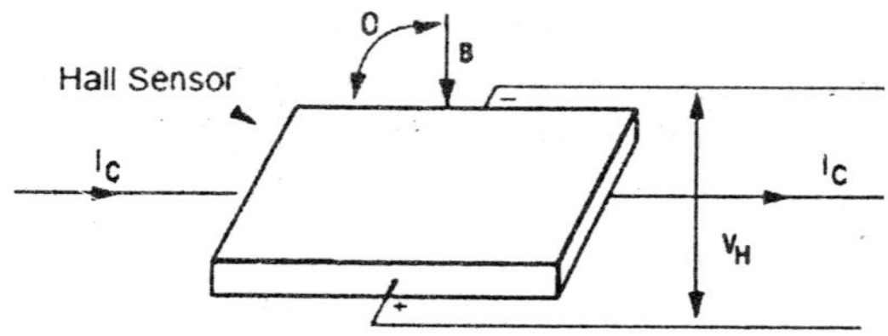

Fig. Hall Sensor

Note: VH = KHOC IC B sin – θ or if sin – θ = (i.e., – θ = 90°)
VH = KHOC IC B or VH = gB B

where:
VH = Hall output voltage, mV
KHOC = gIB (open circuit product sensitivity constant), mV/mA kG
gB = magnetic sensitivity (loaded or unloaded) at a specified control Current, mV/kG
Ic = control current, mA (ac or dc)
B = magnetic flux density, kG (ac or dc)

DISCOVERY OF HALL EFFECT

Edwin Herbert Hall discovered the "Hall effect" in 1879 while working on his doctoral thesis in Physics under the supervision of Professor Henry A. Rollin. Dr. Hall was pursuing the question as to whether the resistance of a coil excited by a current was affected by the presence of a magnet. Through a myriad of experiments and failures, Hall discovered that a magnetic field would skew equipotential lines in a current-carrying conductor.

This effect is observed as a voltage (Hall voltage, VH) perpendicular to the direction of current in the conductor.

Hall conducted an experiment by putting a thin gold leaf on a glass plate and then tapping off the gold leaf at points down its length. He then conducted other experiments using various materials in place of the gold leaf, and various experimental placements of tapping points. In 1880, full details of Hall's experimentation with this phenomenon formed his doctoral thesis and was published in the American Journal of Science and in the Philosophical Magazine.

Kelvin, himself a most distinguished scientist, called Hall's discovery comparable to the greatest ever made by Michael Faraday. The magnitude of this discovery is even more impressive considering how little was known about electricity in Hall's time. The electron, for instance, was not identified until more than 10 years later.

The "Hall effect" remained a laboratory curiosity until the latter half of this century because materials available prior to recent years only produced low levels of Hall voltage. With the advent of semiconductor technology and the development of various III-V compounds, it became possible to produce Hall voltages many orders of magnitude greater than with earlier materials. Thus, semiconductor technology launched the practical design and production of the Hall sensor.

TYPICAL SHAPES AND SIZES

Hall sensors are available in a wide variety of shapes and sizes for adaptability of shapes and sizes for adaptability to many different applications.

The two basic types are transverse and axial. The transverse type is useful where the field must be measured in thin gaps and for multiplier applications. The axial type must be used where the field is parallel to the axis of a hole, such as in traveling wave tubes or solenoids.

Standard transverse probes as thin as 006" and axial probes as small as 063" in diameter are available. Bulk-material Hall plates may be sandwiched between ferrite pieces to obtain effective air gaps less than 003". This may be useful in applications requiring maximum magnetic efficiency, such as electronic compasses and proximity sensors.

For a Hall sensor to accurately measure flux density, the Hall plate area should be smaller than the cross section of the field to be measured. The output voltage is proportional to flux density, but a Hall plate is not equally sensitive over its entire area. If a high resolution is important, the Hall plate area should be small. Active areas as small 0.010" are available, while even smaller ones have been made.

TYPICAL APPLICATIONS

The following are just some of the many applications where Hall *Sensors are used:*

- Magnetic Card Readers
- Proximity Sensors
- Rotary Speed Sensors
- Watt Measurement
- Multipliers
- Magnet Field Measurements
- Electrical Power Measurements
- Current Sensors
- Brushless dc Motors
- Compasses
- Gaussmeters
- Watt-hour Meters
- Permanent Magnet Measurements
- Air Gap Measurements

- Magnetic Circuit design
- Flux Leakage Measurements
- Nondestructive Memory Readouts
- Linear/Angular Transducers
- Magnetic Tape Heads
- Guidance Systems
- Ignition Systems

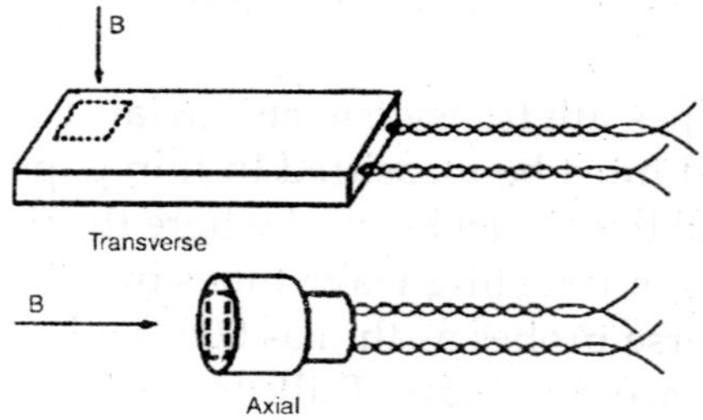

MATERIALS

The Hall effect is basically a majority carrier mechanism depending on the bulk-material properties of the semiconductor material. Unlike transistors and diodes, it is completely independent of surface effects, junction leakage currents and junction threshold voltages.

These factors account for its high stability, reproductivity and reliability when compared to other semi-conductor devices. To obtain a high output voltage the active element must have a high Hall coefficient, RH. Also, since the output is proportional to the current density through the element, its resistance should be as low as practical to prevent excessive heating.

Since the noise output is essentially thermal, low resistance is also an important requirement for devices to be used at very low signal levels. Some of the semiconductor materials used for Hall sensors are indium antimonide (InSb), indium arsenide (InAs) and gallium arsenide (GaAs). GaAs generators have high output and very high resistance making them relatively noisy and the temperature coefficient of the output voltage is less than -0.1%/°C.

InSb has high output and low resistance, but the temperature coefficient of the output voltage is about -1%/°C. InAs has less output than InSb, but its temperature coefficient is less than -0.1%°C and its resistance is also low. These considerations make InAs the most suitable materials for many Hall effect applications.

Indium Arsenide Hall sensors may also be made of deposited thin films.

These units do not exhibit the same low resistance and high mobilities as their bulk-material counterparts, but they do offer advantages which may be realized in many applications. These advantages include lower current requirements for comparable output voltages, and significantly low cost. For those applications where excellent linearity and stability are required, bulk material Hall sensors are recommended.

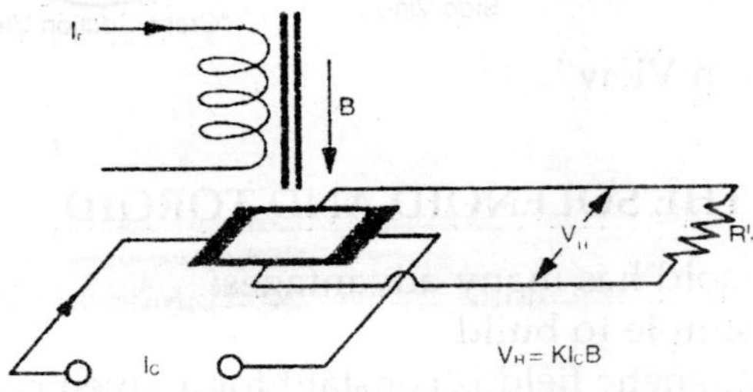

This schematic representation illustrates both the measuring and multiplying capabilities of a Hall sensor. By holding Ic constant, VH becomes a direct function of B, the magnetic flux density. If both Ic and B are variable, VH is proportional to the product of the two functions. Holding B and Ic constant, VH becomes a function of the angle between B and the plane of the Hall-sensor active area.

TOROID (DONUT)

We now consider another important structure, the toroid. A toroid is a solenoid whose ends have been wrapped around to form a donut. Toroid's were used to construct the original computer memory, "core memory." Toroid's with a break are still in use today to record video and audio onto magnetic tapes.

5h15.50 Toroid and iron filings

PURPOSE: To show the field of a toroid.

DESCRIPTION: A toroid mounted in a clear platter is placed on the overhead and powered with a power supply. A clear magnetic compass is used to probe the field.

EQUIPMENT: The toroid is stored in the electricity and magnetism cabinet.

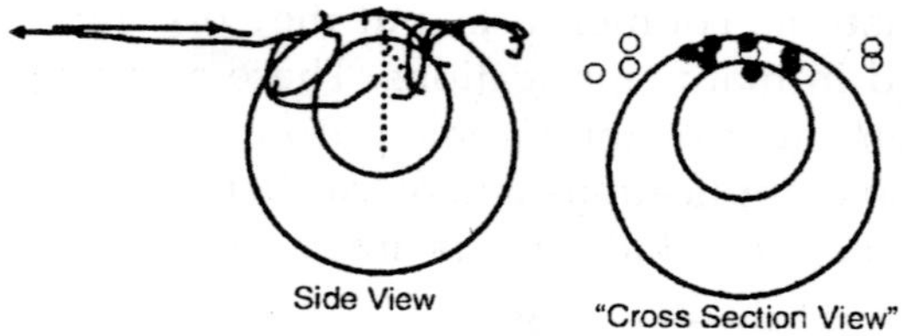

"Cross Section View".
Side View.

COMPARING THE SOLENOID AND TOROID

A. The solenoid has many advantages:
   1. It is simple to build.
   2. It's magnetic field is constant for a given current and number of loops.

B. The toroid has the advantage of limiting the field completely inside the toroid.

C. The problems with the toroid is that
   1. They are more difficult to build than solenoids
   2. Their magnetic field is not uniform since it depends on r.

Chapter 6

Measurements Concerning Electrons

As early as about 1830, electrical discharges in gases were intriguing a number of experimental physicists in Europe. In 1881, at the Cavendish Laboratory at the University of Cambridge, J.J. Thomson began experimenting with gaseous discharges, and continued to do so for the next 50 years. When Thomson started his research, cathode rays had already been known for about 50 years, but their nature was controversial.

Joseph John Thomson was born on December 18, 1856 in Manchester. His father died when he was only 16. Young Thomson attended Owens College in Manchester, where his Professor of Mathematics encouraged him to apply for a scholarship at Trinity College, one of the most prestigious of the colleges at Cambridge University. Thomson finished second in his class in the graduation examination in mathematics in 1880 and won the scholarship. Trinity College gave him a fellowship where he stayed upon. He engaged himself in developing mathematical models that would reveal the nature of atoms and electromagnetic process.

The Cavendish Laboratory at Cambridge was founded in 1871 with James Clerk Maxwell - who developed the basic equations of electromagnetism as the first Cavendish Professor. At 28, the young Thomson was chosen to be the third Cavendish Professor in 1884 following Maxwell and Lord Rayleigh. True, he was inexperienced in doing experiments, but he learned quickly. Supported by his administration and teaching, many important experiments on electromagnetism and atomic particles were performed. Many outstanding Physicists received their early training at the Cavendish, including 7 Nobel Prize winners and 27 Fellows of the Royal Society. Thomson took a keen interest in the work of all the young researchers, daily checking on their progress and making suggestions for improvement.

J.J. Thomson married Rose Paget on January 22, 1890. She was among the researchers at the Cavendish as one of the first generation of women permitted into advanced university studies. She performed experiments of soap films in 1889 after attending some of Thomson's lectures. They had two children: George Paget Thomson who flourished into a prominent Physicist himself and won Nobel Prize for discovery of the diffraction of electron by crystals in 1937. Their daughter John Paget Thomson often accompanied his father in his travels.

In the paper published in Philosophical Magazine 100 years ago in October 1997, Thomson reported that cathode rays were charged particles, which he called "corpuscles". It is hard to recall any discovery since then that has had more impact on not only physics but science, technology and our daily lives. We shall briefly follow the course of history in this article that made it possible.

CATHODE RAYS

Indeed, as far back as 1705, it had been noticed that sparks from an electrical machine would jump further in rarified air than in air at normal pressures. Watson in 1748 observed an aurora borealis like "arch of lambent (i.e. glowing) flame" in a glass tube of rarified air 32 inches long. In 1838 Faraday sent a current from an electrostatic machine through a glass tube containing air at low pressure and observed a purple glow extending from the positive electrode, or anode, at one end, almost to the negative electrode, or cathode, at the other.

The cathode was covered with a glow, and there was a dark space between this glow and the purple column. The dark space has since been called the "Faraday Dark Space." (The colour of the internal glow of such tubes depends on the kind of gas present in the tube. Neon, at pressures approximately one hundredth that of atmospheric pressure, glows with a bright orange colour when current passes through it; helium, a pinkish white; mercury vapour, a light greenish blue.)

Although Faraday observed a number of interesting phenomena, he was limited by the fact that the suction pumps available at the time for reducing the gas pressure were not too efficient. A great step forward was made, about 1854, when H. Geissler, a German glass blower of exceptional skill, not only developed an improved vacuum suction pump, but succeeded in sealing into glass tubes wires attached to metal electrodes.

The evacuated Geissler tubes which he made were particularly suitable for the study of the passage of electricity through gases at low pressure, and with them J. Plucker, in Germany, made numerous experiments between the years 1858 and 1862. Among other things, he observed that the tube in the vicinity of the cathode, i.e. the electrode attached to the negative side of the source of potential, emitted a green glow or luminescence. The position of the glow could be changed by bringing a magnet up to the tube.

The studies of electrical discharge through gases were continued in Germany by Plucker's pupil, W.Hittorf (1869), and by E.Goldstein (1876). From their observations they concluded that the luminescent glow on the tube was caused by "rays" originating at the cathode, which Goldstein consequently called cathode rays. The rays could be deflected by a magnet and were also able to cast a shadow of an obstacle placed in their path, showing that they traveled in straight lines.

Between the years 1879 and 1885 the English scientist William Crookes, who designed improved vacuum dischage tubes, made a very comprehensive series of investigations of the electrical discharge. From these he concluded that the cathode rays actually consisted of a stream of negatively charged particles, which were expelled from the cathode — the negative electrode — with extremely high velocities.

This view of the nature of cathode rays supported a suggestion made in 1872 by C.F.Varley, but it was opposed by many European physicists, including such eminent men as E.Wiedemann (1880), H.Hertz (1883) the discoverer of radio waves, and (1894) P.

Lenard. The latter group thought the cathode rays were an electromagnetic wave motion or vibration, analogous to light waves but of shorter wave length.

If the rays are really a stream of charged particles then they should be deflected by passage through an electric field, as well as by a magnetic field, although Goldstein, in spite of several trials, had failed to observe any such effect. But the deflection of cathode rays in the field of magnet was an accepted fact, and this could not be explained if the rays were similar to light waves.

In an apparently decisive experiment, performed by J. Perrin in France in 1895, the cathode rays were allowed to fall on a device known as a Faraday cylinder, connected to an electrometer by means of which the sign and magnitude of electric charge could be determined.

It was found that a negative charge collected in the cylinder, and so it was argued that the rays were made up of negative particles. Objection was taken to this conclusion on the grounds that negatively charged particles might well be ejected from the cathode, but there was no proof that they are identical with the cathode rays.

Atomic Models: On the basis of his experiments, J.J. Thomson proposed a model of internal atomic structure according to which atoms consisted of a positively charged substance (positive electric fluid) distributed uniformly over the entire body of the atom, with negative electrons embedded in this continuous positive charge like seeds in a watermelon, or raisins in pudding. Since electrons repel each other but are, on the other hand attracted to the centre of the positive charge, they were supposed to assume certain stable positions inside the body of the atom. Ernest Rutherford (1871-1937) and Hans Geiger together bombard thin pieces of gold with alpha particles.

Most of the alpha particles passed right through the foil, and the result was exactly what the experimenters expected based on Thomson's model of the atom. But some of alpha particles struck the gold foil and were deflected at a sharp angle often 90° or more (J J Thomson's Model).

This amazed Rutherford, who remarked "It was as though you have fired a 15-inch shell at a piece of tissue paper and it came back and hit you". Early in 1911 Rutherford exclaimed to Geiger, "I know what the atoms looks like!". Rutherford put together a new idea of the atom: what if all the positively charged particles in the atom were not spread like a fluid throughout the atom as Thomson had thought but were lumped together in the centre in one tiny area, or "nucleus"?

Most of the atom's mass would be contained in the nucleus, and an equal number of negatively charged electrons would be found in motion somewhere outside the nucleus. Undoubtedly, it was a compelling idea - a sort of tiny planetary system that resembled the larger solar system we are living in.

The required proof was provided in 1897 by J.J. Thomson, the famous English physicist, whose work has had a profound effect, both direct and indirect, on the study of atomic structure.

In the first place, he repeated Perrin's experiment and confirmed that charged particles are emitted by the cathode. But, in addition, he showed that when the cathode rays are deflected by a magnetic field, as indicated by the change in position of the luminescence they produce, the negatively charged particles are correspondingly deflected.

Further, Thomson succeeded, where Goldstein and others had failed, in deflecting the path of the cathode rays by means of an electric field. Previous failures had been due to excessive ionization of the gas still present in the discharge tube, thus offsetting the effect of the electric field. By working at very low pressures Thomson minimized the influence of this ionization and then he was able to observe the anticipated deflection. We shall trace his efforts that established that the cathode rays are actually a stream of particles carrying negative electrical charges. Indeed, this was the result of a number of converging studies by several prominent physicists, which we shall briefly consider.

The Experiments: First, in a variation of an 1895 experiment by Jean Perrin, Thomson built a cathode ray tube ending in a pair of metal cylinders with a slit in them.

These cylinders were in turn connected to an electrometer, a device for catching and measuring electrical charge. Perrin had found that cathode rays deposited an electric charge. Thomson wanted to see if, by bending the rays with a magnet, he could separate the charge from the rays.

He found that when the rays entered the slit in the cylinders, the electrometer measured a large amount of negative charge. The electrometer did not register much electric charge if the rays were bent so they would not enter the slit. As Thomson saw it, the negative charge and the cathode rays must somehow be stuck together: you cannot separate the charge from the rays.

All attempts had failed when physicists tried to bend cathode rays with an electric field. Now Thomson thought of a new approach.

A charged particle will normally curve as it moves through an electric field, but not if it is surrounded by a conductor, say, a sheath of copper. Thomson suspected that the traces of gas remaining in the tube were being turned into an electrical conductor by the cathode rays themselves. To test this idea, he took great pains to extract nearly all of the gas from a tube, and found that now the cathode rays did bend in an electric field after all.

Thomson's Experiments

Physicist J.J. Thomson at Cambridge carried out a series of definitive experiments on cathode rays around the year 1897. By turning them slightly off course with electrical forces, as shown in the figure, he showed that they were indeed electrically charged, which was strong evidence that they were material. Not only that, but he proved that they had mass, and measured the ratio of their mass to their charge, m/q.

Since their mass was not zero, he concluded that they were a form of matter, and presumably made up of a stream of microscopic, negatively charged particles. When Millikan published his results fourteen years later, it was reasonable to assume that the charge of one such particle equaled minus one fundamental charge, q=-e, and from the combination of Thomson's and Millikan's results one could therefore determine the mass of a single cathode ray particle.

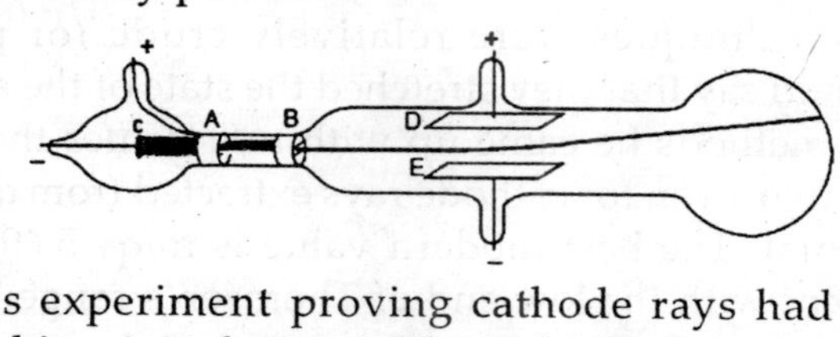

Thomson's experiment proving cathode rays had electric charge (redrawn from his original paper). The cathode, c, and anode, A, are as in any cathode ray tube. The rays pass through a slit in the anode, and a second slit, B, is interposed in order to make the beam thinner and eliminate rays that were not going straight. Charging plates D and E shows that cathode rays have charge: they are attracted toward the positive plate D and repelled by the negative plate E.

The basic technique for determining m/q was simply to measure the angle through which the charged plates bent the beam. The electric force acting on a cathode ray particle while it was between the plates would be proportional to its charge,

$$F_{elec} = (\text{know constant}).\, q$$

Application of Newton's second law, a=F/m, would allow m/q to be determined:

$$\frac{m}{q} = \frac{\text{know constant}}{a}$$

There was just one catch. Thomson needed to know the cathode ray particles' velocity in order to figure out their acceleration. At that point, however, nobody had even an educated guess as to the speed of the cathode rays produced in a given vacuum tube. The beam appeared to leap across the vacuum tube practically instantaneously, so it was no simple matter of timing it with a stopwatch!

Thomson's clever solution was to observe the effect of both electric and magnetic forces on the beam. The magnetic force exerted by a particular magnet would depend on both the cathode ray's charge and its velocity:

$$F_{mag} = (\text{known constant\#2}).qv$$

Thomson played with the electric and magnetic forces until either one would produce an equal effect on the beam, allowing him to solve for the velocity,

$$v = \frac{(\text{known constant})}{(\text{know constant\#2})}$$

Knowing the velocity (which was on the order of 10% of the speed of light for his setup), he was able to find the acceleration and thus the mass to charge ratio m/q.

Thomson's techniques were relatively crude (or perhaps more charitably we could say that they stretched the state of the art of the time), so with various methods he came up with m/q values that ranged over about a factor of two, even for cathode rays extracted from a cathode made of a single material. The best modern value is m/q= 5.69×10^{-12} kg/C, which is consistent with the low end of Thomson's range.

The Cathode Ray as a Subatomic Particle: The Electron

What was significant about Thomson's experiment was not the actual numerical value of m/q, however, so much as the fact that, combined with Millikan's value of the fundamental charge, it gave a mass for the cathode ray particles that was thousands of times smaller than the mass of even the lightest atoms.

Even without Millikan's results, which were 14 years in the future, Thomson recognized that the cathode rays' m/q was thousands of times smaller than the m/q ratios that had been measured for electrically charged atoms in chemical solutions.

He correctly interpreted this as evidence that the cathode rays were smaller building blocks - he called them electrons - out of which atoms themselves were formed. This was an extremely radical claim, coming at a time when atoms had not yet been proven to exist! Even those who used the word "atom" often considered them no more than mathematical abstractions, not literal objects.

Thomson concluded from these two experiments, "I can see no escape from the conclusion that [cathode rays] are charges of negative electricity carried by particles of matter". But, he continued, "What are these particles? Are they atoms, or molecules, or matter in a still finer state of subdivision?" This is the famous "duck argument". If it looks like a duck, quacks like a duck and waddles like a duck, then we have good reason to believe it is a duck!

Thomson's third experiment sought to determine the basic properties of the particles. Although he couldn't measure directly the mass or the electric charge of such a particle, he could measure how much the rays were bent by a magnetic field, and how much energy they carried. From this data he could calculate the ratio of the mass of a particle to its electric charge (m/e). He collected data using a variety of tubes and using different gases.

The method essentially involved sending the beam into an electrically shielded collector, as in the Perrin's experiment, but in this case making the collector physically small. The beam gave up its charge to the collector and also heated it by mechanical impact.

The quantity of heat energy, H, given to the collector in a given interval of time T could be determined from its mass, specific heat, and temperature rise. This charge in temperature could be measured by means of a very light thermocouple attached to the collector. The total charge, Q delivered to the collector could be measured by a sensitive electrometer.

Electron As A Particle In 1899, Thomson set out to resolve the doubt concerning the significance of the e/m values of the "corpuscles" by determining directly their charge as well as the charge to mass ratio. Unfortunately, this could not be done with the cathode-ray particles, and so he turned to another source. It was well-known towards the end of the

19th century that ultraviolet light falling on certain metals, particularly zinc, was associated with the emission of negatively charged particles, a phenomena known as the photoelectric effect.

Thomson determined the en ratio for these particles by means of electric and magnetic fields and found it to be virtually the same as for the cathode ray corpuscles. Charged particles emitted by an incandescent filament, i.e. by the thermo-ionic effect, also had a similar e/m value. His estimate of the electronic charge of the photoelectric particles turned out to be similar to the unit electronic charge.

In view of the consistency of the e/m for the negatively charged particles produced in different ways, it was reasonable to conclude that the particles were identical. In the words of Thomson: "The experiments just described, taken in conjunction with previous ones.... On cathode rays, show that in gases at low pressures negative electrification, though it may be produced by very different means, is made up of units each having a charge of electricity of a definite size; the magnitude of this negative charge is.... equal to the positive charge carried by the hydrogen atom (ion) in the electrolysis of solutions".

Because the charge on the particles present in the cathode rays, and associated with the thermo-ionic and photoelectric effect, was identical with the elementary electric charge, the name electron originally intended by G. Johnstone Stoney for magnitude of the charge, soon became associated with the actual particles themselves.

Assuming that n particles each of mass m and velocity v hit the collector in time T, and that each particle carries a charge q, then

$$Q = nq$$

provided each particle "sticks" to the collector, and hence deposits its charge to be measured, and does not cause any secondary emission of charged particles. Thomson tried to ensure that these conditions would be met by grounding his shielding electrode. (Even so, if the collector acquires a negative potential from the incoming beam, then it is possible that some of the beam may bounce away from it and be collected by the shield.)

Further, if each particle is collected it will relinquish its kinetic energy to the collector, and this energy will be evident in the form of heat:

$$H = n(1/2mv^2)$$

Dividing Equation (1) by Equation (2) one finds

$$q/m = 2Qv^2H$$

It is hence necessary to measure velocity v.

Now, in a magnetic field oriented so as to be perpendicular to the original path of the rays, the resulting path is observed to be part of a circle. The magnetic field, then, must be exerting a centripetal force upon the ray particles. Assuming that each of the particles has a mass m, a velocity v, and charge q, and that they move in a magnetic field of intensity B in a path of radius of curvature R, one can write the following equations:

Magnetic force on particle = Centripetal force for circular motion

$$Bqv = mv^2/R$$

Which can be rearranged to give

$$q/m = v/BR$$

Hence, combined with magnetic field deflections, the relations (3) and (5) would give values for both q/m and v. Using this method, Thomson found values of v of the order of 2.4 × 107 to 3.2x107 meters per second (about one tenth the velocity of light), and from 1.0 to 1.4x1011 coulombs/ kilogram for the ratio of charge to mass for cathode ray particles.

In the same paper that described the method described above, Thomson reported a different method for getting the needed second relationship between q/m and v.

In this second method he used a tube popularly known as the Thomson tube. The schematic diagram for determination of q/m is shown elsewhere. The cathode ray beam could be sent through an electric field produced by the plates A and B in which region there could also be a magnetic field perpendicular to the paper established by external coils. Any deflection of the beam could be measured by the scale S at the end of the tube.

Electron As A Wave: The diffraction and interference properties of radiation necessitate a wave structure, but photoelectric phenomena and the Compton effect imply that radiation consists of particles rather than waves.

In Compton effect, an instant X-ray is scattered by a free electron just like in a collision between two rigid spheres. In other words, radiation may be regarded as exhibiting a dual wave - particle behaviour; some of the properties of the radiation may be wave properties where as others are particle properties.

By means of Planck's Quantum Theory equation and the mass energy relationship of Einstein, Prince Louis-Victor Pierre Raymond de Broglie (1892-1987) deduced that a particle mass m moving with a velocity v should be associated with waves of length l (lamda), given by l = h/mv where h is the Planck's constant.

It was calculated that with a moderately high velocity such as could be obtained by passage of an electron through a potential of about 100 to 1000 volts, the de Broglie waves should have a wavelength of the order of 10-8 cms. If this were the case, then crystals should be capable of producing diffraction effects with electrons.

The first definite proof that electrons can be diffracted and consequently exhibit wave, as well as the familiar particle, properties was obtained in the Bell Telephone Laboratories in New York by C.J. Davisson and L.H. Germer in 1927. By studying the reflection and scattering by a nickel crystal, of a beam of electrons, given a specific velocity by passage through a known potential difference, it was found that the electrons behaved like waves rather than particles.

Using electrons which had been accelerated by a potential of 54 volts, the experimental results were found to be equivalent to those expected from radiation of wavelength 1.65 Å, in remarkably good agreement with the value of 1.67 Å calculated by means of the de Broglie equation.

Further, evidence for the existence of electron waves was obtained independently in 1927, by George Paget Thomson (1892-1975), son of J.J. Thomson.

He passed a stream of fast moving electrons through a very thin sheet of metal and then allowed the resulting beam to fall on a photographic plate.

Upon development, the plate showed a diffraction pattern consisting of a series of concentric circles, just as might have been produced by X-rays, indicating that the electrons were manifesting wave properties. With no electrical field applied, the magnetic field can deflect the beam upward or downward as shown, for example, by the dotted path in the figure.

The beam travels in a straight line except in the region of the magnetic field, where its path is an arc of a circle of radius R, as in Equation 5. Neglecting the fringe effects, it is fairly easy to calculate R from the measured deflection of the beam and the geometrical constants of the tube.

Thus this tube may be used, as other tubes had been, for measurements with the magnetic field alone, and under these circumstances Equation 5 would apply. But suppose that while the magnetic field is present, a potential difference V is applied to the two deflecting plates. If the plates are a distance D apart, then the resulting electric field strength between the plates will be $E = V/D$ (6) (If V is measured in volts and D is measured in meters, then E will be found in newtons per coulomb).

A charge q in this field experiences a force Eq, up or down depending on the sign of q and the direction of the field E.

If we arrange the applied potential difference to have a value such that the force E_q is numerically equal, but opposite in direction, to the force on the particles due to the magnetic field (which is $F_{mag} = B_{qv}$, in which v is the velocity of the particles), we would then have

$$F_{elec} = F_{mag}$$

$$\text{or } E_q = B_{qv} \text{ and hence (7)}$$

$$v = E/B \text{ (8)}$$

(It is easy to show that if E is measured in newtons/coulomb, and B in webers per square meter, then v is in meters per second). In practice, the potential V is varied until the beam is observed to be at a position of no net deflection. Zero deflection implies that the electric and magnetic forces are equal. V is then measured with a voltmeter, and B is determined by use of a search coil and ballistic galvanometer. One can then put the numerical value of v thus found back into Equation 5 to determine the value of q/m.

Thomson found, in a series of experiments, values for q/m which, when averaged, came to 0.77×10^{11} coulombs/kilogram. This value disagreed with the one he published from his heating effect experiments, a disagreement he attributed primarily to possible systematic errors in the latter experiments.

In his writings for the next few years he usually gave the value of q/m as "approximately 10^{11} coulombs/kilogram". Thomson boldly announced the hypothesis that "we have in the cathode rays matter in a new state, a state in which the subdivision of matter is carried very much further than in the ordinary gaseous state: a state in which all matter is of one and the same kind; this matter being the substance from which all the chemical elements are built up".

Thomson remarked that this surprising result might be due to the smallness of m or to the bigness of e. He argued that m was small, citing Philipp Lenard, who had shown that the range of cathode rays in air (half a centimeter) was far larger than the mean free path of molecules (10^{-5} cm).

Lenard was awarded the Nobel Prize in Physics in 1905 for studying the cathode rays. If the cathode ray travels so much farther than a molecule before colliding with an air molecule, it must be very much smaller than a molecule. Thomson concluded that these negatively charged particles were also constituents of atoms. From 1897 onward, thanks largely to the experiments of Perrin and Thomson, the corpuscular model for cathode rays received general consent.

Thomson's view that the cathode ray particles were the fundamental building block, or even a fundamental building block of atoms was not, however, received with much enthusiasm. Several other lines of research, notably in the fields of analysis of spectra and of radioactive phenomena, had to converge before the real role of Thomson's corpuscles within the atom could be understood and generally accepted.

Thomson's achievements were honoured in numerous ways, and mark him as among the most accomplished physicists of his era. In 1906 he was awarded the Nobel prize in physics for his researches into the discharge of electricity in gases. In 1918 he was chosen Master of his old college, Trinity, and the next year he resigned the Cavendish Professorship. He guided Trinity with his usual common sense and benevolence until shortly before his death in August 30, 1940. "e" by Millikan's method Millikan's apparatus consisted of two horizontal metal plates about 22 cm diameter and 1.6 cm apart as indicated by A and B. The plates were supported in a closed vessel containing air at low pressure, and were connected to the poles of a high voltage (10,000 volts) battery, V. In the upper plate, there were a number of small holes as represented by C. By means of a atomizer, a fine spray of a non-volatile oil was introduced into the vessel.

As a result of friction in the atomizer, the droplets of oil so obtained were electrically charged. From time to time, one of these droplets would pass through the hole C, and then it could be observed by means of a telescope.

By using the illumination of a powerful beam of light, entering the window W (at left), the droplet appeared as a bright star on a dark background. With the battery V disconnected, the droplet fell slowly under the influence of gravity, and the rate of fall was measured.

This rate (or velocity) represented by v1 is dependent on the mass m of the droplet and is given by the equation v_1 = kmg where g is the gravitational acceleration (981 cm/sec^2) and k is a proporationality constant which is related to the viscosity of the air and the size of the oil droplet. The high voltage battery was then switched on, thus producing the electric field, the direction being such as to make the charged droplet move upward, against the force of gravity.

If E is the strength of the electric field, i.e., the voltage of the battery divided by the distance between the plates, then the upward force acting on the droplet is Een, where en is the charge carried by the droplet. Since this is opposed by the gravitational force mg, the net upward force is Een - mg. The upward velocity v_2 of the oil droplet which is measured, is then represented by

$$v_2 = k\,(Een - mg)$$

The proportionality constant k has the same significance as in the previous equation.

From the above two equations, it is easy to show that

$$en = mg\,(v_1 + v_2)/Ev_1$$

Since the quantities v1, v2, g are available, it is possible to calculate the charge en carried by the oil drop if the mass m were known. Using Stokes Law, applicable to small spherical drops falling under the influence of gravity, it could be shown that

$$v1 = 2gr2d/9h$$

Here h is the density of the oil of which the drops are made. Since v1 has been determined, as described above and g, h and d may be regarded as known the radius r of the drop could be determined from the above equation.

It is now easy to determine the mass of the oil drop which is given by m = 4pr3d/3, inserting this result into the equation for en, together with the measured velocity, v1 and v2, the magnitude of the charge en carried by the oil droplet can now be determined. As a result of a large number of measurements, Millikan found that the charge en was always an integral i.e. a whole number, multiple of a definite elementary charge, which was presumably the electronic charge.

After applying numerous corrections to the foregoing equations, Millikan concluded in 1917 that the most reliable value of the unit charge was 4.774×10^{-10} esu which is very close to the modern value 4.803207×10^{-10} esu.

Millikan and his Oil Drops

Robert Andrews Millikan (1868-1953) was the son of Silas Franklin Millikan a congregational preacher, and Mary James Andrews, a graduate of Oberlin who had been Dean of Women at a small college in Michigan. Raised in Maquoketa, Iowa, where his family moved in 1875, young Millikan enjoyed a story book, Midwestern American boyhood, fishing, farming, fooling and learning next to nothing about science.

In 1886, he enrolled in the preparatory department of Oberlin College and in 1887, in the classical course of the college itself. At the end of his sophomore year, he was asked to teach an introductory physics class. Millikan plunged into the subject, liked it and soon decided to make it his career.

Millikan graduated from Oberlin in 1891 and continued to teach physics to the preparatory students. He was awarded an M.A. for his achievement of successfully pursuing a course of instruction in Dynamic Electric Machinery in 1893.

Millikan entered Columbia University on a fellowship as the sole graduate student in physics. He was impressed by the experimental deftness of Michelson, under whom he studied at the University of Chicago in the summer of 1894.

After receiving his Ph.D. in 1895, Millikan went to Europe for post-graduate study. He heard PoincarŠ lecture at Paris, took a course from Planck at Berlin, and did research with Nernst at Gottingen. In 1896, the excitement of the discovery of X-ray still fresh in his mind Millikan joined the faculty of the University of Chicago as an assistant in physics.

This is where he met Greta Irwin Blanchard, the daughter of a successful manufacturer from Illinois whom he married in 1902. He spend a large fraction of his energies into the development of the physics curriculum, especially in introductory courses and wrote/co-authored a variety of text books which quickly became standards and sold in large numbers. Mainly because of his outstanding pedagogical achievements, Millikan was promoted to an Associate Professorship.

By 1909 Millikan was deeply involved in an attempt to measure the electronic charge. No one had yet obtained a reliable value for this fundamental constant, and some any-atomistic continental physicist were insisted that it was not the constant of a unique particle but a statistical average of a diverse electrical energy.

His famous experiment of determination of the electronic charge is described elsewhere in the article. Of and on all the while Millikan had continued his exploration of the photoelectric effect and by 1950 had confirmed the validity of the Einstein's equation of photoelectric effect in every detail.

Millikan held many important posts and membership of several eminent Academies and Societies. In 1921, Millikan accepted appointment as the Chairman of the Executive Council and Director of the Norman Bridge Laboratory at the California Institute of Technology.

Employing the photonic interpretation of cosmic rays, Millikan developed a theory of their origin in 1928. To find a measure of cosmic ray energies, he put Carl Anderson, a young research fellow at Caltech to work with a cloud chamber set in powerful magnetic field which ultimately lead to which detection of the negative electron - also called Positron, in 1932.

How the "Positive Electron" or the "Positron" was Discovered: The English mathematical physicist P.A.M. Dirac (1902-1984) in 1928 presented theoretical arguments indicating that a particle similar in mass to the electron but carrying a positive charge may exist. His discussion based on relativistic wave mechanics was of a highly abstruse character. However, the proof of the existence of the long-sought positive electron was obtained by C.D. Anderson (1905-1991) at the California Institute of Technology in 1932.

In order to study the so-called cosmic rays, which appears to come from outer space, Anderson in conjunction with R.A. Millikan, had constructed an apparatus known as a cloud chamber which was placed in a very strong magnetic field. In this cloud chamber, the path of an electrically charged particle could be rendered visible and also photographed.

A cloud chamber is based on the fact that whenever an electrically charged fast-moving particle passes through the air (or any other gas), it produces ionization along its tract. If the air through which these particles pass is saturated with water vapour, the ions serve as the centres of condensation for tiny water droplets, and we see long thin tracks of fog stretching along the particle's trajectories.

The intensity of the track provided information concerning the mass of the particle, and the direction in which it was bent in the magnetic field indicated whether the charge was positive or negative. Numerous tracks were observed due to charged particles resulting from the impact on matter of the very highly energetic cosmic rays. A lead plate of 6 mm thickness was placed across the chamber with the object of depriving the particles of some of their energies.

Anderson stated in one of his lectures: "The degree of curvature in the magnetic field shows a difference depending on the amount of energy lost in the plate. Measurements made on the track of a particle before and after it has passed through the plate, together with observations of the density of the track itself, give definite information about the mass of the particle and the magnitude of the electric charge it carries" The photograph shows one of the numerous photographs obtained in this manner - a photograph of historical significance, for its interpretation by Anderson lead to the discovery of the positive electron. Since the curvature of the track is less below the plate than above, the energy of the particle is greater below the plate.

Hence the particle must have been moving upward. Knowing the direction of the magnetic field and the direction of the motion of the particle, the curvature of the track to the left immediately showed that the particle must be positively charged.

The density of the track was less than would be expected for a proton, but its length was greater. "Photographs of these positively charged particles could be understood only if the particles were assumed to have a mass approximately equal to that of the ordinary electron of negative electric charge, and thus the first evidence for the existence of the positive electron.... was obtained", Anderson said.

Other cloud chamber photograph examined in the light of new discoveries provided further proof that positive electron were produced by the action of cosmic rays. Some photographs showed charged particles to fall into two groups, one being deflected in one direction and the other in the opposite direction by the magnetic field, representing negative electrons and the positive electrons respectively. Anderson suggested the name "Positron" for the positive electron, and this immediately became into general use.

Millikan was an able populariser and lecturer and after he won the Nobel Prize in 1923, he became perhaps the most famous American scientist his days. He was an outspoken, religious modernist. Even after his retirement from his professorship in 1946, he remained active as a public lecturer and spoken frequently on the subjects of science and religion.

By the time of his death, he had been awarded numerous medals even from honorary degrees and professorship of 21 foreign scientific Societies, including the Royal Society of London and Institute de France.

Thomson did not use the term "electron" to refer to his negatively charged particles; he preferred the term "corpuscle". "Electron" had been introduced by the Irish physicist G.Johnstone Stoney in 1891, as the name of the "natural unit of electricity", the amount of electricity that must pass through a solution to liberate one atom of hydrogen. Stoney did not associate the electron with a material particle, and physicists at the time questioned whether or not electricity might be a continuous homogeneous fluid.

The early determinations of the charge of the electron had not established that there was a fundamental unit of electricity. That was because the experiments measured the total charge of a cloud of droplets, without showing that the value obtained was anything other than a statistical average. The same was true for Thomson's measurement of e/m for a beam of cathode rays.

It was the experimental work of Robert Millikan at the University of Chicago, beginning in 1909, that provided the next step in establishing the electron as a fundamental particle. Millikan not only demonstrated that there was a fundamental unit of electrical charge; he also measured it accurately.

Millikan's experimental apparatus and the method he used for the determination of the electronic charge is described in a box. He allowed single oil drops to fall a known distance in air, and measured the duration of the fall. He then turned on an electric field and measured the time it look for each drop to travel the same distance upward. (The oil drops were travelling at constant terminal velocity). These two time measurements let him determine both the mass of the drop and its total charge.

The charge on the oil drop sometimes changed spontaneously, by ionization or absorption of charge from the air. Millikan also induced such changes with either a radioactive source or x-radiation. One could calculate the change in the charge on a drop and the changes in that charge were small integral multiples of e, a fundamental unit of charge.

Millikan wrote, "The total number of changes which we have observed would be between one and two thousand, and in not one single instance has there been any change which did not represent the advent upon the drop of one definite, invariable quantity of electricity or a very small multiple of that quantity". Millikan's final value for e was (4.774 ± 0.009) × 10^{-10} esu. (The modern value is 4.803 207 × 10^{-10} esu).

Millikan associated his measured e both with the charge on Thomson's corpuscles and the charge on the hydrogen ion in electrolysis. He combined his value for e with contemporary measurements of e/m by electrolytic and cathode ray techniques to determine that the mass of Thomson's corpuscle was 1/1845 that of the hydrogen atom surprisingly close to 1/1837.15, the modern value.

Now one had both a definite mass and a definite charge for this would be fundamental particle, and it behaved exactly as one would expect a negatively charged particle to behave. There was now good evidence for believing that it was a constituent of atoms in other words, the electron. He was awarded the Nobel Prize for his work in 1923.

Electron Gets Established

1. *The Zeeman Effect: After* Hertz had shown experimentally that electromagnetic waves are, in fact, produced by oscillating electric charges as predicted by Maxwell's equations, it became commonly agreed that light waves were electromagnetic waves and that they were due to some sort of oscillation of charged particles within, or associated with, molecules or atoms. In 1896 a Dutch physicist, Pieter Zeeman tried to see whether an external magnetic field would affect the wavelength of the light given out by these hypothetical oscillators. His apparatus was, in principle, quite simple: a light source (for example, sodium vapour in a gas flame) was placed between the pole faces of an electromagnet, and the light from the source was sent through a spectroscope. In his attempts Zeeman found that the spectral lines were not changed or shifted when he switched on the magnet. He gave up the experiment, but then happened to read Faraday's accounts of his final experiments some forty years before. He found that Faraday had tried essentially the same experiment. Zeeman's admiration for his predecessor was so great that he decided that if Faraday had thought the experiment worth doing, then he, Zeeman, ought to be willing to put in a little extra effort to repeat it. With a somewhat stronger magnetic field he found that the spectral lines (he was using the well known "D lines" of sodium vapour) were slightly broadened. The broadening was of the order of one fortieth of the separation between the two lines, or about 0.15 Angstrom Unit (Angstrom unit = 10^{-10} metre).

A few years later, it turned out that if one assumes that the emission of light takes place from small charged particles revolving in orbits in the atoms, then one can predict a slight contraction or expansion of the orbits when an external magnetic field is applied. The expansion or contraction of the orbits would result in a slight shift in the wavelength of the emitted electromagnetic radiation. The actual amount of the shift can be predicted if one knows the ratio of q to m for the orbiting particles and the strength of the magnetic field. Lorentz and Zeeman were thus able to postulate the presence within the atom of small charged particles, and to estimate from the line broadening that the ratio of q to m for these particles would have to be about 107 emu per gram. This was, as they were quick to point out, the same ratio that Thomson had found for his cathode ray particles. Lorentz and Zeeman were jointly awarded the Nobel Prize in 1902 for their work.

2. *The Photoelectric Effect:* Hertz in his famous 1887 experiments with electromagnetic radiation had made what seemed like an incidental or an accidental discovery, which was that ultraviolet light falling on certain metals caused them to emit negatively charged particles, or what is known as the photoelectric effect.

The discovery of the electron at once suggested the hypothesis that the photoelectric effect is due to the liberation from the illuminated metal plate of electron which under the influence of the electric field pass from cathode to anode, thereby causing photoelectric current.

This hypothesis was confirmed by Lenard who showed that the photoelectric discharge is deflected in a magnetic field exactly as are cathode rays. By measuring the deflection of the "photoelectric rays" in a known magnetic field, he found a value of e/m about 1.2x107 in qualitative agreement with Thomson's value of e/m for electrons.

In 1899 Thomson applied the technique that he had used with photoelectrically emitted particles to a determination of q/m for the negatively charged particles that, as Edison (the same Thomas Alva Edison) had discovered, are emitted by white hot metals. Thomson found q/m for these particles to be 0.87×10^{11} coulombs per kilogram, again in satisfactory agreement with his value for cathode ray particles. In the next few years Owen in England and Wehnelt in Germany found similar values for particles emitted by certain metallic oxides heated to a red heat.

Lenard found out in 1902 that there was no relationship between the intensity of the light and the energy of the electrons emitted. And a brighter light might cause more electrons to be emitted, but they would not be any more energetic than those released by a dim light. Classical physics could offer no explanation.

By the turn of the century it was known that certain radioactive materials emitted negatively charged particles that had come to be called "beta rays". In 1900 Becquerel sent a beam of such particles through electric and magnetic fields to determine their velocity and their ratio of charge to mass. He found the then rather astonishing velocity of approximately 2/3 that of light, and a ratio of charge to mass of about 10^{11} coulombs per kilogram. Kaufmann, in 1901 and 1902, determined q/m for beta rays more precisely, finding it to be 1.77 × 1011 coulombs per kilogram.

That's where Einstein stepped in, breaking out Planck's quantum theory, which had been gathering dust for a couple of years without too much attention. Planck had pointed out that light emits distinct "packets"; Einstein added that light also travels in packets. Einstein pointed out that a particular wavelength of light is made up of quanta of fixed energy content, according to quantum theory. When a quantum of energy bombards an atom of a metal, the atom releases an electron of fixed energy content and no other.

A brighter light would contain more quanta, still always of fixed energy content, causing the emission of more electrons, also still all of the same energy content. The shorter the light's wavelength (and the higher the frequency), the more energy contained in the quanta and the more energetic the electrons released. Very long wavelengths (of lower frequency) would be made up of quanta having much smaller energy content, in some cases too small to cause any electrons to be released. And this threshold would vary depending on the metal.

This was the first use of Planck's theory since its invention to explain the blackbody problem - and once again it succeeded in explaining a physical phenomenon where classical physics could not. For this work, Einstein received the 1921 Nobel Prize in Physics. It was the first major step in establishing what would become known as quantum mechanics, the recognition of the discrete and discontinuous nature of all matter, especially noticeable on the scale of the very small.

All paths lead to Rome: Thus within four or five years of Thomson's 1897 investigations, he and others were able to show that electrons, as they were then commonly called, with essentially the same properties are emitted from all sorts of materials by several different mechanisms:

(a) By strong electric fields, or by bombardment of the cathode by positive ions, as in the classical cathode ray tube.
(b) As a result of absorption of ultraviolet light by atoms.
(c) By thermal agitation of atoms in white hot metals or oxides.
(d) By some spontaneous process within radioactive atoms.

In addition, the Zeeman effect could best be interpreted as showing that precisely similar particles exist within atoms. (This is of some significance, because it is not logically necessary for an atom to "contain" some particle that it is later observed to emit. Modern physics abounds with examples of emitted particles that are produced, as it were, on the occasion of their emission).

Thus by 1900 the electron was well established as a constituent of atoms. Already physicists were working toward a better knowledge of the electron's inherent characteristics, its charge, mass, and size, and toward an understanding of its role in an astonishingly wide array of chemical and physical phenomena. But follow the story further we must.

Electron through the 20th Century: In 1913, not long after Millikan's oil drop results, Niels Bohr constructed a theory whose confirmations provided support for the view that the electron was both a fundamental particle and a constituent of atoms. Bohr developed his theory based on Rutherford's nuclear model of the atom put forward in 1911 (in contrast to his mentor Thomson's idea that the positively charged particles in the atom were spread like a fluid throughout the atom a sort of plum pudding!).

Rutherford's model of the atom had a small, massive positively charged nucleus orbited by electron of mass in and charge -e. A decade later, Bohr's theory was superseded by the Quantum Mechanics of Erwin Schrodinger and Werner Heisenberg which also assumed an electron with charge e and mass m and it gives exactly the some predictions as the Bohr theory for the Balmer series in the Hydrogen atom.

In the 1920, the experiments of Otto Stern and Walther Gerlach established the existence of the spatial quantizations which provided the evidence for an intrinsic spin of the electron, that is, it behaves as though like a tiny spinning top and a magnet of certain strength.

The electron itself has turned out to be not quite the creature that J.J. Thomson thought it was. According to the quantum theory developed by Albert Einstein and others, it is a mistake to think that electrons must be either particles or waves but not both. Under some conditions electrons act like particles; under other conditions they act like waves. (The wave character of electrons was in fact experimentally indicated by J.J. Thomson's own son, G.P. Thomson, who as a result shared the Nobel Prize in 1937). Physicists have also found that electrons are only the most common members of a whole "family" of related fundamental particles -- all of them infinitesimal points carrying charge, mass, and something called "spin". Why the particles have these properties remains a mystery, a grand challenge for the next century of research.

In some experiments, the charge-to-mass ratio is the only quantity that can be measured directly. Often, the charge can be inferred from theoretical considerations, so that the charge-to-mass ratio provides a way to calculate the mass of a particle. The electron charge to mass quotient, *e/m*, is a quantity in experimental physics. It matters because the electron mass m is difficult to measure directly, and is instead derived from measurements of the fundamental charge e and e/m. It also has historical significance: Thomson's measurement of *e/m* convinced him that cathode rays were particles, which we know as electrons.

The 2006 CODATA recommended value is: e/m = *1.758820150(44)* C/kg.

In 1897 J. J. Thomson, an English physicist, conducted a series of experiments on cathode rays. As a result of his experiments, Thomson was able to measure the charge to mass ratio of the electron; he could not however, measure accurately the charge or mass independently. The measurement of the electron's charge independently was achieved by Millikan by his famous experiment from 1909 and with Thomson's results also a value for the electron mass was obtained. This experiment is called the oil-drop experiment and it was the first successful scientific attempt to detect and measure the effect of an individual subatomic particle.

Robert Andrews Millikan was born on the 22nd of March, 1868, in Morrison, Ill. (U.S.A.), as the second son of the Reverend Silas Franklin Millikan and Mary Jane Andrews. As a scientist, Millikan made numerous momentous discoveries, chiefly in the fields of electricity, optics, and molecular physics. His earliest major success was the accurate determination of the charge carried by an electron, using the elegant "falling-drop method"; he also proved that this quantity was a constant for all electrons, thus demonstrating the atomic structure of electricity. Next, he verified experimentally Einstein's all-important photoelectric equation, and made the first direct photoelectric determination of Planck's constant h (1912-1915).

He has been the recipient of the Comstock Prize of the National Academy of Sciences, of the Edison Medal of the American Institute of Electrical Engineers, of the Hughes Medal of the Royal Society of Great Britain, and of the Nobel Prize for Physics 1923.

He died on the 19th of December, 1953, in San Marino, California.

We shall use the next formula to calculate the mass of the electron in Bohr's atom:

$$m = \frac{\hbar}{4\pi r^2 Rc}$$

and where are:

$R = 1.0973731568527 * 10^{7}\ m^{-1}$, Rydberg_constant

$r = 0.52917720859 * 10^{-10}$ m, Bohr_radius

$h = 1.005457162884 * 10^{-24}$ Js, Dirac's_constant

c = 2.99792458*108m/s, speed_of_light_in_vacuum

Using the best experimental values for r,R,h_bar and (the value of c is fixed by definition) gives:

$$m = 9.109382149955857*10^{-31}kg$$

Now, we shall use the another formula to calculate again the mass of the electron in Bohr's atom:

$$m = \frac{4\pi R\hbar}{\alpha^2 c}$$

where the value of fine structure constant is:

$$\alpha = 7.297352537641*10^{-3}$$

Using the best experimental values for r,R,h bar and (the value of c is fixed by definition) gives the value electron mass:

$$m = 9.109382149956873 10^{-31}\ kg$$

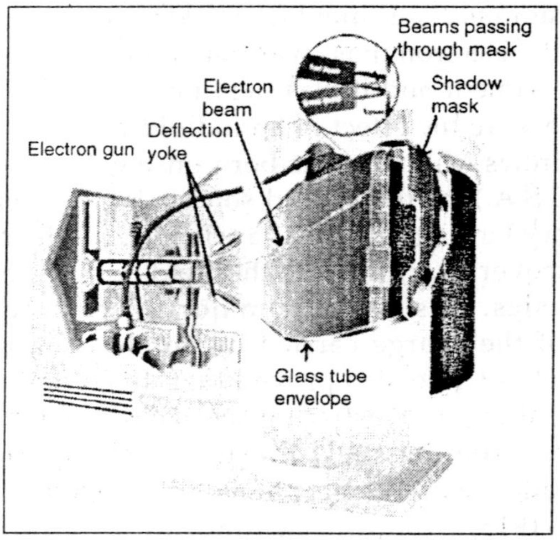

Electron mass (electron rest mass) is atomic fundamental physical constant used as atomic unit of mass, m = *9.109 382 15 (45)* * 10^{-31} kg.

The 2007 Rasko Jovanovic recommended value is:

$$m = 9.10938214995 * 10^{-31}\ kg$$

CATHODE RAY TUBES (CRTS)

(a) Most common display device today
(b) Evacuated glass tube
(c) Heating element (filament)
(d) Electrons attracted to focusing anode cylinder
(e) Vertical and Horizontal deflection plates
(f) Beam strikes phosphor coating on front of tube.

CRTs, or video monitors, are the most common output device on computers today.

The basic structure of a CRT. A CRT is an evacuated glass tube, with a heating element on one end and a phosphor coated screen on the other. When a current flows through this heating element, called a filament, the conductivity of the metal filament is reduced due to the high temperature.

These cause electrons to pile up on the filament, because they can not move as fast as they would like to. Some of these electrons actually boil off of the filament. These free electrons are attracted to a strong positive charge from the outer surface of the focusing anode cylinder (sometimes called an electrostatic lens).

However, the inside of the cylinder has a weaker negative charge. Thus when the electrons head toward the anode they are forced into a beam and accelerated by the repulsion of the inner cylinder walls in just the way that water is speeds up when its flow though a smaller diameter pipe. By the time the electrons get out they're going so fast that they fly past the cathode they were heading for.

The next thing that the electrons run into are two sets of weakly charged deflection plates. These plates have opposite charges, one positive the other negative.

While their charge is not strong enough to capture the fast moving electrons they do influence the path of the beam. The first set displaces the beam up and down, and the second displaces the beam left and right. The electrons are sent flying out of the neck of the bottle, until they smash into the phosphor coating on the other end of the bottle.

The impact of this collision on the out valence bands of the phosphor compounds knocks some of the electrons to jump into the another band. This causes a few photons to be generated, and results in our seeing a spot on the CRT's face.

LIQUID CRYSTAL DISPLAYS (LCDS)

Currently, the most popular alternative to the CRT is the Liquid Crystal Display

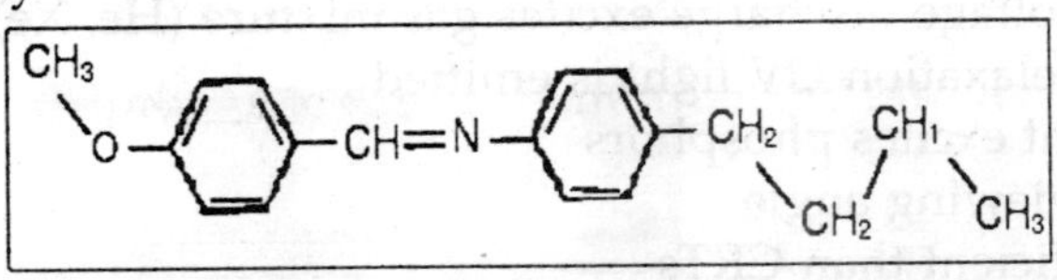

(LCD). LCDs are organic molecules that, in the absence of external forces, tend to align themselves in crystalline structures. But, when an external force is applied they will rearrange themselves as if they were a liquid. Some liquid crystals respond to heat (i.e. mood rings), others respond to electromagnetic forces.

When used as optical (light) modulators LCDs change polarization rather than transparency (at least this is true for the most popular type of

LCD called Supertwisted Nematic Liquid crystals). In their unexcited or crystalline state the LCDs rotate the polarization of light by 90 degrees. In the presence of an electric field, LCDs the small electrostatic charges of the molecules align with the impinging E field.

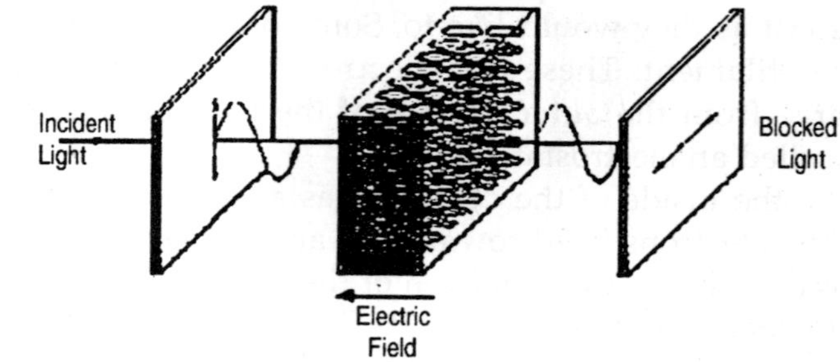

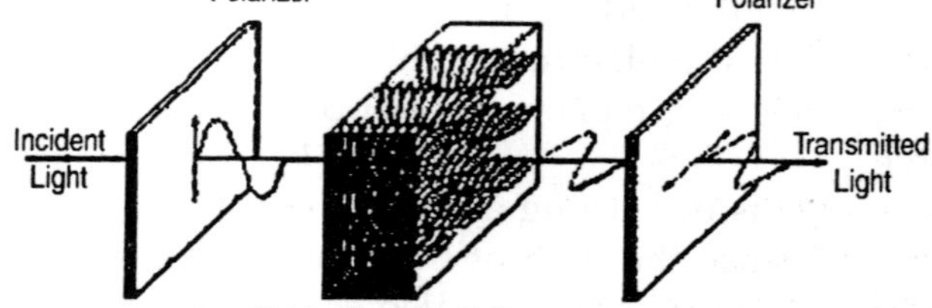

The LCD's transition between crystalline and liquid states is a slow process. This has both good and bad side effects. LCDs, like phosphors, remain "on" for some time after the E field is applied. Thus the image is persistent like a CRT's, but this lasts just until the crystals can realign themselves, thus they must be constantly refreshed, again, like a CRT.

PLASMA DISPLAY PANELS

a) Promising for large format displays
b) Basically fluorescent tubes
c) High-voltage discharge excites gas mixture (He, Xe)
d) Upon relaxation UV light is emitted
e) UV light excites phosphors
f) Large viewing angle
g) Less efficient than CRTs
h) Not as bright
i) More power
j) Large pixels (~1mm compared to 0.2 mm for CRT)
k) Ion bombardment depletes phosphors

Plasma display panels (PDPs) are essentially a matrix of very small fluorescent tubes with red, green, and blue phosphors. As in ordinary tubes, a discharge is initiated by a high voltage which excites a mixture of inert-gases such as He and Xe. Upon relaxation, ultraviolet (UV) radiation is generated which excites the phosphors.

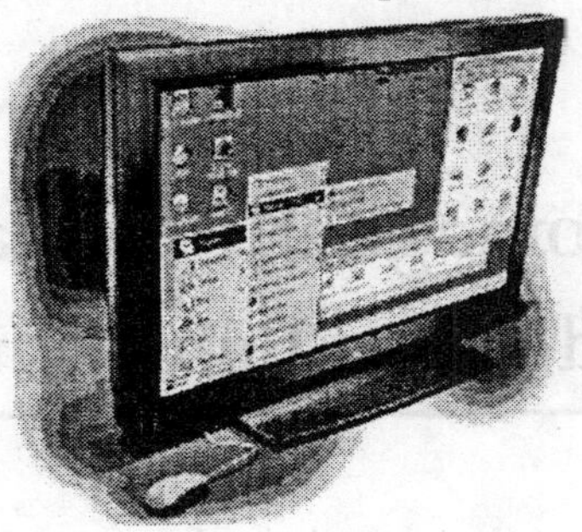

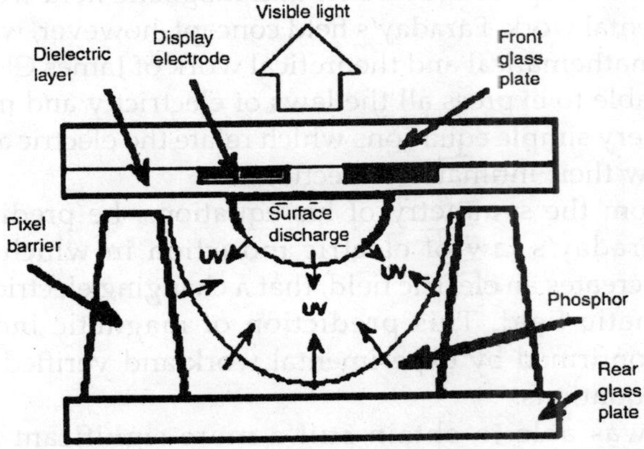

PDPs provide a large viewing angle since the phosphors emit light uniformly. A 40-inch PDP typically consumes about 300 W whereas the peak brightness is only 1/3 of that of a CRT consuming about half the power. Sealing and vacuum pressure support problems apply to PDPs as well, requiring thicker glass as the screen is enlarged.

In addition, the discharge chambers have pixel pitches of more than 1 mm which makes it difficult to construct high-definition television (HDTV) and workstation monitors. By contrast, TFTLCDs, CRTs and FEDs may have pixel pitches as small as 0.2 mm.

Chapter 7

Electromagnetic Radiation and Wave Behaviour

Faraday's concept of the electric and magnetic field was a great aid to his experimental work. Faraday's field concept, however, was absolutely crucial to the mathematical and theoretical work of James Clerk Maxwell. Maxwell was able to express all the laws of electricity and magnetism in terms of four very simple equations which relate the electric and magnetic fields and show their intimate connection.

In fact, from the symmetry of his equations, he predicted that, in analogy to Faraday's law of electric induction in which a changing magnetic field creates an electric field, that a changing electric field would create a magnetic field. This prediction of magnetic induction was immediately confirmed by experimental work and verified the validity of Maxwell's equations.

Maxwell was able to obtain still a more significant insight into electromagnetic phenomenon from the study of his equations. He discovered the existence of a solution to his equations in which there is an absence of charge and in which the electric and magnetic fields behave like a wave. He associated this solution with the phenomenon of light which he recognized as electromagnetic radiation. As a result of this insight, he was able to explain the emission, absorption and propagation of light.

The concept of an electromagnetic wave is easy to understand once its relation to electric and magnetic induction is realized. Consider an electric field oscillating at some point in space. Then, by magnetic induction, (or since a changing electric field produces a magnetic field) the oscillating electric field creates a magnetic field perpendicular to itself in its immediate neighborhood.

This oscillating magnetic field, as a result of electric induction, then creates an oscillating electric field which, in turn, induces an oscillating magnetic field which, in turn, induces an oscillating electric field and so on and so forth. In this way, an electromagnetic wave propogates through empty space at the velocity of light which is 186,000 miles per second or 3×10^8 meters per second.

The production of electromagnetic radiation can be achieved by causing a charged particle to oscillate back and forth since this causes the electric field associated with the charged particle to also oscillate. This is precisely how radio waves, another form of electromagnetic radiation, are produced in a radio antenns.

A current of electrons is made to oscillate up and down in the antenna at a given frequency in order to broadcast radio waves. The absorption of electromagnetic radiation occurs as a result of charged particles interacting with the oscillating electric and magnetic fields of the radiation. For example, the eye detects visible light when the electrons in the retina become activated by the electric and magnetic fields of the light ray.

Electromagnetic radiation comes in a variety of different forms such as microwaves, radio waves, infra-red (heat) radiation, visible light, ultraviolet radiation, X rays and gamma rays. All of these forms of electromagnetic radiation are identical in the sense that they are oscillating electric and magnetic fields which all propagate at the velocity of light. They differ only in that each one represents a different range of frequencies and hence, wavelengths. The frequency of a wave is the number of times per second that the electric and magnetic fields oscillate back and forth. The wavelength is the distance between successive maximums of the fields and is inversely proportional to the frequency. A list of the frequency and wavelength of the various forms of electromagnetic radiation is given in the accompanying table.

The range of frequencies of visible light and the range of frequencies that the sun radiates with maximum intensity exactly overlap. Apparently, the eye has been biologically adapted so as to detect the electromagnetic radiation of the sun. If human life had developed on a planet orbiting a star which emitted principally infra-red radiation then, through the process of evolution, one would expect visible light for these people to be in the infra-red range. The visual detection of objects using infra-red radiation has actually been achieved by the military. They have developed a photographic film sensitive to infra-red rays which they have used for aerial photo reconnaissance at night.

In addition to perceiving electromagnetic radiation, they is also sensitive to the different frequencies in the visible light range. This accounts for our colour vision. Each colour corresponds to a different range of frequency in the visible range as is listed in the above table. The order of the colours in the list is exactly the same as the rainbow. This is no coincidence. The sun radiates visible light of all frequencies and hence, all colours. White light is merely a combination of all the colours. When white light or sunlight propagates through water such as a raindrop or glass such as a prism, the light ray is bent both upon entering and leaving the medium of water or glass. The amount of bending a ray of light experiences depends on the frequency of the ray; the greater the frequency, the more it is bent. The different colours that compose light are therefore separated when they propagate through a medium like water or glass and hence, one observes a rainbow.

The knowledge that sunlight is composed of all the colours helps us to understand why the sky is blue and why sunsets and sunrises are red. When one looks up at the sky on a cloudless day, one observes blue light. This light is sunlight which has been absorbed by air molecules in the upper atmosphere and reradiated towards earth.

Because the amount of light reradiated at any one particular frequency is proportional to the fourth power of the frequency, more blue light than red is reradiated towards earth and hence, the sky appears blue. At sunset or sunrise, light arriving from the sun has a thicker envelope of air to travel through in order to reach us as is illustrated above. Since more blue light than red has been absorbed out of the beam of sunlight as it travels through the atmosphere, the light reaching us during a sunset or sunrise appears red.

Maxwell's identification of light with oscillations of the electric and magnetic fields explains the wave nature of light. Long before Maxwell's identification, it was realized that light behaves as a wave. The wave nature of light was first suggested by Christian Huyghens, a contemporary of Newton. In fact, he and Newton had a long standing controversy concerning the nature of light. Newton adopted the position that light was a beam of particles and hence, could not display wave behaviour. Huyghens had a difficult time convincing the scientific world of the wave nature of light because of the formidable reputation of his scientific foe. After the results of a number of experiments corroborating Huyghens point of view became known, however, the science community finally adopted the wave picture of light.

By an ironic twist of fate, however, experiments performed in the past century have revealed that, although light behaves in many situations as a wave, there are instances when it also behaves like a beam of particles. So, there is a sense in which Newton was correct. However, from the point of view of the experimental evidence that was available to Huyghens and Newton, it was Huyghens who made the more accurate interpretation of the data.

We will defer our discussion of the wave-particle duality of light to the time when we discuss atomic physics, and turn our attention instead to the wave nature of light. Let us first consider the nature of wave behaviour in general by discussing a more familiar example, namely, the waves of the sea. When one looks at the surface of the ocean on a windy day, one sees alternate rows of crests and troughs as are schematically shown in the above figure.

As one observes the movement of the water, one observes that the crests and troughs are moving toward the shore. One might conclude that the water is moving towards the shore but, in fact, the water composing the wave is actually moving up and down.

It is oscillating in the direction perpendicular to the direction in which the wave is moving. This can be easily verified by watching a buoy bobbing up and down in the water. At the shore, it is true that at certain moments water moves toward the shore, but it is also true that, at other moments, an equal amount of water moves away from the shore as the wave washes back into the sea.

One should not confuse the two different types of motion one encounters in wave behaviour. One motion is the motion of the wave or really, the wave form which is continuous and unidirectional. The other motion is the actual movement of the medium which is always an oscillatory motion. In the example of waves upon the ocean, the oscillatory motion of the medium is perpendicular to the motion of the wave form. This type of wave is called a transverse wave and is differentiated from a longitudinal wave in which the medium oscillates back and forth in the same direction in which the wave moves.

Perhaps the best known example of a longitudinal wave is a sound wave. A sound wave requires the existence of transmitting medium. Most of the sound waves with which we come in contact propagate through the atmosphere although sound waves can also propagate through solids and liquids. There are no sound waves on the surface of the moon however, because there is no atmosphere.

Let us consider the production and propagation of sound waves in our atmosphere. A sound wave is produced as a result of the rapid vibration of some object, like a violin string for example, which causes the molecules of air surrounding it to move back and forth like the vibrating string.

This causes alternate condensations and rarefications of the air molecules as is illustrated in the accompanying diagram. As the string moves to the right, it pushes the air molecules together creating a condensation; as it moves back to the other direction, it leaves a rarefication.

The motion of the air molecules back and forth is in the same direction as the motion of the vibrating string. The vibrating column of air adjacent to the string will, in turn, cause the column of air adjacent to it to vibrate and then, this column of air will act on the column adjacent to it and so on and so forth and in this way, the sound wave will propagate through the air. Each column of air will vibrate back and forth and hence, there will be no net displacement of the air as the sound wave propagates from the vibrating string to the ears of some listener.

The air molecules that are contact with violin strings will not come in contact with the listener's ear. The vibration of these air molecules will propagate through the air to the listener's ear, however, as a consequence of contact the air molecules make with each other through collisions. The vibrations of the final column of air adjacent to the eardrum of the listener will cause his eardrum to vibrate with the same frequency of the original violin string.

The vibrations of the eardrum are transmitted through tiny bones to a cavity containing a fluid where they activate nerve cells which transmit the information to the brain. The human ear is capable of detecting frequencies in the range from sixteen vibrations per second to twenty thousand.

The greater the frequency of the sound wave, the higher the pitch that we detect. The loudness of the sound wave depends on the strength of the vibrations or on the distance through which the string vibrates. The harder the string is struck, the greater is the amplitude of its vibration and hence, the louder the sound. The frequency of the string does not depend on the strength with which it is struck but rather on the length of the string, its thickness and the amount of tension with which it is strung.

We have now considered both transverse waves, (waves on the ocean) and longitudinal waves (sound waves). In both cases, the wave is transmitted as the result of oscillatory motion of a physical medium. In the case of ocean waves, the water was oscillating up and down in the direction transverse to the propagation of the wave.

In the case of the sound wave, the air molecules were oscillating back and forth in the direction longitudinal to the wave motion. We encounter a somewhat different situation when we consider electromagnetic radiation since no medium is required to propagate the wave motion. Instead of the oscillation of some physical medium, electromagnetic waves involve the oscillation of the electric and magnetic fields. The oscillation of these fields is transverse to the direction of the wave propagation and hence, light is a transverse wave.

We are still left with the mystery of how a wave is able to propagate through empty space. This mystery is related to the mystery of action at a distance discussed earlier in connection with electric and magnetic forces. The solution to these two related mysteries, provided by Faraday and Maxwell, is the concept of a field. Later in this book, after we have studied more about elementary particles and their basic interactions, we shall return to this mysterious question and consider another possible solution.

Although it is more difficult to conceive the wave nature of light than that of the ocean because of the absence of a concrete medium, light, nevertheless, displays exactly the same wave behaviour characteristic of waves on the water. Let us consider two phenomena characteristic of waves, namely, linear super-position and interference. If I drop a stone into a quiet pond of water, a wave front with a circular shape will propagate from the point where the stone enters the water. If two stones are dropped into the water a short distance apart, two circular waveforms will propagate.

The two waves will flow through each other without affecting each other, that is, after passing each other the two waves are exactly the same as they were before, i.e., they retain their circular form. In the region where they meet, however, they interfere with each other.

The motion of the water up and down due to the two waves will either add together or subtract depending on whether or not two crests arrived at the same point or a crest and a trough. If two crests arrive at the same place, then, the waves add such that a crest of twice the height of a single wave is created.

If, on the other hand, the crest from one wave arrives at the trough of another, then, the two waves can momentarily cancel so that it appears there is no disturbance of the water at all at this point. However, an instant later, as the two waves propagate past each other, one observes the two waves again. The two cases of constructive and destructive interferences are illustrated for a linear wave of one pulse in the figure below.

Light also can interfere constructively or destructively with itself just like water waves. Let us consider light from the same source shining through two slits of some opaque material. This situation is analogous to the dropping of two stones since spherical light waves will emanate from each of the two slits.

If we now observe the light from these two slits projected on a screen, we will observe a pattern of alternating illuminated and dark patches. Those positions which are illuminated are the places where the light from the two sources arrived in phase and the dark points where the two beams of light arrived out of phase. It was the two slit interference experiment first performed in 1789 by Thomas Young which finally settled the controversy between Newton and Huyghens concerning the wave or particle nature of light.

Chapter 8

Electricity and Magnetism

FUNDAMENTALS OF ELECTRICITY AND MAGNETISM

A Newtonian description of nature requires an understanding of the basic forces that operate between the various constituents of the universe. So far, the only force we have discussed in any detail is the gravitational force.

We shall now turn our attention to the electric and magnetic forces, which, we will soon discover, are intimately connected to each other and are referred to collectively as the electromagnetic interaction. The electromagnetic interaction plays a major role in determining both the structure of matter and the general interactions of matter. The reason for this is, that of the three major constituents of matter, the proton, the neutron and the electron, two of them, the electron and proton, are charged and hence, exert an electric force whereas all three possess magnetic moments and hence, exert magnetic forces.

As a result of this the structure of atoms and molecules, as our future studies will reveal, are governed by the electromagnetic interaction. For example, the electrons of an atom are held in their orbits about the nucleus by the electric force.

Not only are the forces inside the atom and molecule electromagnetic, but also the forces between atoms and molecules are electromagnetic. Molecular bonds are responsible for the structure of gross matter. Hence, the resistance one feels when one tries to penetrate solid matter is electromagnetic as is the force exerted by a coiled spring. The forces produced by chemical action are also electromagnetic since all chemical reactions are governed by the electric properties of an atom's outer electrons. The production of light, as we will soon see, is also a result of the electromagnetic interaction.

In addition to the natural phenomenon referred to, we encounter the effects of the electromagnetic interaction in a steadily increasing number of devices and machines invented by man for his pleasure such as the light bulb, the neon sign, the radio, the television, the computer and the laser, not to mention the electric streetcar, the electric stove, the electric dishwasher, the electric iron, the electric broom, the electric can opener, etc. etc. etc.

Let us begin our study of the electromagnetic interaction by turning our attention to the electric force which, in some ways, is like the gravitational force. The strength of the electric force, like the gravitational force, is inversely proportional to the square of the distance between the two interacting bodies.

In addition to the inverse square law the force in both cases acts along the line joining the two bodies and is equal and opposite for the two bodies as is illustrated in the accompanying diagram.

The electric force differs from the gravitational force, however, in two vital ways. First of all, the gravitational force is only attractive whereas the electric force can be both attractive and repulsive depending on the signs of the charges involved. A charged particle is either positive like a proton or negative like an electron. Charges of the same sign repulse each other whereas charges with opposite signs attract each other.

The other difference in the two forces is the fact that the strength of the electric force is considerably greater than that of the gravitational force. In fact, the electric attraction of a proton and an electron is 1040 times stronger than their gravitational attraction. This accounts for the very strong forces which hold the atom together, create the chemical bonds in molecules and produces the molecular bonds in solids and liquids.

Because of the fact that the electron and proton are exactly equal and opposite in charge, gross matter is electrically neutral. In fact, if there was a slight difference in the magnitudes of the electron's and proton's charge of only one part in 1020 macroscopic matter would be completely unstable for the repulsive forces generated by such a minute difference in charge would be enough to completely destroy all the molecular bonds which hold matter together and scatter it to the four corners of the universe.

Fortunately for our existence the equality of charge for the proton and electron, two particles which differ in so many ways, is apparently identical. Macroscopic matter becomes electrically charged, however, whenever there is a slight excess or shortage of electrons. This occurs, for example, when a rubber rod is stroked by a piece of cat fur in which case electrons are transferred by friction from the fur to the rubber rod.

The rubber rod has an excess of electrons and hence, has a net negative charge whereas the fur contains more protons than electrons and hence, has a net positive charge. If, after rubbing the rod with the fur, one were to put these two objects in contact, electrons attracted by the positive charge of the fur would flow from the rod to the fur until electric neutrality was once again established.

Electric neutrality can also be established by placing a copper wire between the rod and the fur which would permit the passage of an electric current of flow of electrons.

Not all materials, however, permit the passage of an electric current. Some materials such as wood, asbestos and rubber, referred to as insulators, do not permit the flow of an electric current because all of their electrons are tightly bound by the chemical bonds holding the material together. In certain materials, however, such as metals, not all of the electrons are so tightly bound.

These materials, referred to as conductors, permit the flow of an electric current. When electric current flows, electrons do not flow from one end of the wire to the other, the electrons in a conductor behave more or less like the molecules in a gas.

When there is no current flowing in the conductor, the free electrons move back and forth within the wire in a random fashion colliding with each other. When a current is flowing there is a general drift of the electron in a particular direction.

The net effect is that a current flows from one end of the wire to the other although no actual electron makes this trip. The wire is actually electrically neutral with as many electrons flowing into any one segment as flow out of that segment. The heat generated by an electric current is due to the collisions of the electrons in the conductor. These collisions are also responsible for the resistance of the conductor. This explains why, for a given current, the amount of heat generated in a conductor is proportional to its resistance.

Rubbing a rubber rod with a piece of cat's fur and placing a wire between them is not the most efficient way of creating an electric current. However, the more efficient methods of producing an electric current are based on the same idea of creating two polarities of charge which a flow of current will neutralize.

This is the principle of a battery with the difference that the separation of the positive and negative charge is a perpetual process brought about by a chemical reaction. A battery is simply two different metal bars placed in an acid solution. When the acid acts upon the metal, it combines chemically with the metal depositing an excess of electrons on the rod. Since the acid works on one of the metal rods of the battery faster than the other, more electrons build on one of the metal rods than the other.

If a wire is placed between the two metal rods of the battery an electric current will flow in order to neutralize the excess of electrons. However, this flow of electrons activates the battery to continue acting chemically on the metal rods to produce an excess of charge which causes a current flow and so on until the acid finally eats away one of the rods and then the battery can no longer generate an electric current. From the study of electric currents generated by devices like the battery, it was discovered that electric currents exert a force on each other. This force can not be attributed to the electric force.

Although an electric current involves the flow of electrons, the current in a wire is neutral. As stated earlier, the current is due to the drift of the free electron in the copper wire in one particular direction. This direction is actually opposite to the flow of the current.

This apparent backward way of defining the current is due to the historical accident that Benjamin Franklin defined a current as a flow of positive charge at least one hundred years before the electron was actually discovered. Although there is a current due to the drift of the electrons, the wire itself is electrically neutral. If one examines a given section of a wire, as much charge flows out of it as in it. The force observed between wires carrying currents is therefore not due to the electric force but some other force.

It was also discovered that an electric current exerts a force on a magnetic compass. In fact, one can make an artificial magnet by wrapping copper wire around an iron bar and passing a current through the wire. It is clear from these two experiments that an electric current behaves like a magnet and that the force between two currents is magnetic.

Further study showed that the magnetic force, like the electric force, is also inversely proportional to the square of the distance between the two currents. Since the current in a wire has a net charge of zero, one can not ascribe the magnetic force solely to the charge of the particles within. It is clear that it is the motion of the charge particles that produces this new force which differs from the electric force in a number of ways.

Perhaps the best way of comparing the two forces is to consider the interaction of two positively charged particles moving parallel to each other. There is a repulsive electric force between these two charges whose strength is proportional to the product of their charges divided by the square of the distance between.

In addition to this electric force there is also an attractive magnetic force whose strength is also proportional to the product of the charges divided by the square of the distance between them. The magnetic force is also proportional, however, to the product of their velocities divided by the velocity of light squared. Since the velocity of a particle can never be greater than the velocity of the light, the magnetic force is always less than the electric force.

The force depends on the relative direction of the two currents in a complicated fashion. When the two currents are parallel (anti-parallel) the force is attractive (repulsive). The force acts along the line connecting the two currents and is equal and opposite. For other configurations, the force is not always directed along the line connecting the two currents and it is not always equal and opposite.

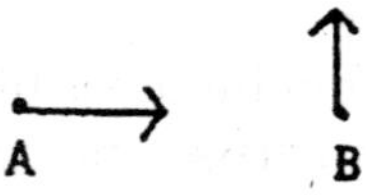

Consider the two positively charged particles moving perpendicular to each other as indicated by the arrows in the above diagram. Particle B exerts a magnetic force on particle acting in the same direction that B is moving whereas A exerts no force on B at all. It is clear that the magnetic force is a good deal more complicated than the electric force.

The connection between the magnetic properties of an electric current and a lodestone or magnet is easily made by considering the atomic structure of a magnet. All atoms, because of their electrons orbiting the nucleus, have equivalent electric currents which can exert magnetic forces. Since the orientation of atoms in matter is so completely random, the effects of each individual atom's magnetism cancel. In certain very select materials, such as lodestones, the atoms are oriented in such a way that the magnetic forces exerted by individual atoms can add up constructively to create a rather strong magnetic force.

This explains why a magnet loses its magnetism if it is dropped or heated since, in both of these cases, the special orientation of the magnet's atoms are destroyed. This also explains why the north pole of one magnet attracts the south pole of another since it is in this position that the internal currents of the atoms are parallel and hence, attractive. When one of the magnets is rotated so that now two north poles are facing each other or two south poles, then the internal currents are anti-parallel and the two magnets repel each other.

Both the electric and magnetic interactions of charged particles discussed above have a magical quality about them in the sense that the charged particles interact with each other at a distance without any apparent physical connection between them. With the exception of the gravitational force, which also has this magical property of action at a distance, all the other forces between bodies require some kind of physics contact. The concept of action at a distance is very difficult to comprehend. Try conceiving of how you personally could move some object without coming into physical contact with it. This is the work of a magician. Yet every proton, every electron, is a magician since they excerpt forces on each other through a vacuum with absolutely nothing between them. How can one account for this?

Michael Faraday invented the concept of the electric and magnetic field in an attempt to understand this mystery. According to his idea, each charged particle created an electric field about it. If the particle is in motion, then, in addition to the electric field, it also creates a magnetic field. This electric and magnetic field spread through all space. The field, at any given point in space, is inversely proportional to the square of the distance from the charged particle generating the field. Some charged particle finding itself in the electric field generated by another charged particle will experience and electric force according to the strength of the electric field and its own charge. If this particle is in motion then it will experience a magnetic force proportional to its charge, its velocity and the strength of the magnetic field.

While the concept of the electric and magnetic field might solve the mystery of action at a distance, it leaves us with a new mystery to wit; how does a charged particle create an electric or magnetic field at some distance from it without a connecting medium. Faraday and his successor actually believed in the existence of an invisible medium which they called aether which, according to them, filled up all of space.

This concept survived until the advent of the theory of relativity when Einstein showed that the existence of an aether was inconsistent with experimental facts. The electric and magnetic field, as far as I am concerned, do not really exist in the same sense that charged particles exist. The field concept is an abstraction, a pictorial description of the electric and magnetic forces. It is, nevertheless, a conceptual construction which is very useful. The concept of electric and magnetic fields will help us to understand the phenomenon of electrical induction. It will also help to explain the creation, absorption and propagation of light as an electromagnetic phenomenon.

The phenomenon of electrical induction was first discovered by Faraday, who observed that an electric current momentarily within a loop of wire into which a magnet has been thrust as is illustrated in the above diagram. The current flows only while the magnet is in motion either being inserted into the loop or being withdrawn from it. It was not just the presence of the magnet but its motion that induced an electric current. Faraday discovered that a current could also be induced by moving the loop of wire with respect to the magnet. In other words, it is the relative motion of the magnet and the loop which induces a current. If the current flows in a clockwise direction when the magnet is inserted, then it flows counter-clockwise when the magnet is removed. Furthermore, if the opposite pole of the magnet is inserted into the loop, then the current also changes direction.

An understanding of the mechanism of induction can be made by recalling that it is the relative motion of the magnet and the loop of wire which causes a current to flow. Let us consider a loop of wire moving with respect to the magnet. The motion of the loop creates an effective current out of every electron in the wire. The direction of this current is not along the wire but perpendicular to it i.e. in the direction of its motion. Each of these individual currents is acted upon by the magnetic field of the magnet which exerts a force perpendicular to the direction of motion. It is this force which causes the current to flow in the looped wire. The same effect occurs when the magnet is moving and the wire is at rest since it is only the relative motion of the two which matters.

Electrical induction also operates with two loops of wire facing each other. If an electric current flowing in the first loop changes in any way, it causes a momentary current to flow in the second loop. The principle is the same as that of electrical induction caused by a moving magnet. When the current in the first loop of wire changes, the magnetic field generated by this current changes at the position of the second loop. In the case of induction involving the relative motion of the loop of wire and the magnet, the magnetic field at the loop of wire is also changing.

It is clear from these two cases that the induced electrical current is caused by a changing magnetic field. Since the induced electrical current is caused by an electric field, the phenomenon of induction can be expressed totally in terms of fields. A changing magnetic field induces an electric field. Faraday's law states that the strength of the induced electrical field produced is proportional to the rate of change of the magnetic field and acts perpendicular to the direction in which the magnetic field is changing.

PRELUDE TO RELATIVITY

With Maxwell's description of light as electromagnetic radiation, the physicists of the nineteenth century felt that they had a complete description of the physical universe. Perhaps there were certain details to be studied but they believed that all of the basics were understood. In actuality, they were on the eve of a revolution in physics that would shake the foundations of their thought as much as the Copernican revolution had shook the thought of their predecessors some three hundred years earlier. Their entire notion of space and time would change as a result of this revolution which Einstein's Theory of Relativity would bring. But before discussing the nature of this revolution, let us first consider the experimental observations that lead to the theory of relativity.

Maxwell had shown through his equations describing the electric and magnetic fields that light was an electromagnetic wave which traveled at the velocity c, equal to 2.998 cm./sec. (or 186,000 miles per sec.). The first measurement of the velocity of light was made by Roemer in 1676 making use of astronomical observations of the eclipses of the Jovian moons by Jupiter itself. The satellite L_o orbits Jupiter every 42 hours. Roemer noticed that the time interval between successive reappearences of L_o varied depending on the position of the earth. When the earth travels along its orbit about the sun from position A to B, the time interval between successive reappearances of L_o increases.

This is due to the fact that the earth is farther from Jupiter after each successive reappearance and an extra amount of time is required for the light reflected from the moon to reach the earth. As the earth travels back from position B to A, the time intervals for successive reappearances of L_o decreases. The overall time lag as the earth travels from position A to B is 22 minutes which is the time re quired for the light to travel the diameter of the earth's orbit.

Using the value of the diameter of the earth's orbit about the sun, Roemer calculated the velocity of light, c, to be 2.14 x 10^{10} cm./sec. This value is fairly close to the presently known value of 3 x 10^{10} cm/sec. The chief source of error in his calculation was due to the poor estimate in his day of the distance between the earth and the sun. Although Roemer's determination of c was not very precise, one can not help but be impressed by the ingenuity of his approach.

A more accurate determination of c was made in 1725 by James Bradley, who also made use of astronomical observations. He observed that the stars that lie above the plane of the earth's orbit seem to move about a very tiny circle in the sky in the course of a year. This phenomenon, known as the aberration of starlight, is due to the fact that the angle at which the star is observed changes as the earth moves about its orbit and hence, the star appears to also move in a circle. The angle at which the star is observed depends on the velocity of the earth. In order for the starlight coming directly from overhead to enter the telescope and subsequently be observed. Bradley was able to determine the velocity of light. He found the value 3.1 x 10^{10} cm./sec., a value extremely close to the modern value.

More accurate measurements of c were performed without recourse to astronomical observations using terrestrial-based apparatus which utilize either a rotating mirror or else a rotating cogwheel. A measurement by Michelson using a rotating mirror involved an optical path of 22 miles between Mt. Wilson and Mt. San Antonio in California in 1880.

One of the triumphs of Maxwell's equations was that the velocity of the electromagnetic waves that he predicted was in agreement with the measured value of the velocity of light. Maxwell's identification of light as electromagnetic radiation explained a great deal about the nature of light. One mystery still remained, however, namely, how is it that t light wave can propogate through empty space? It was impossible for physicists to conceive of a wave without some medium in which it travels. One need only consider the wave motion of water or sound waves to see that these waves could not exist without their medium. Using these analogies, Maxwell postulated the existence of transmitting medium for light which he called the luminiferous aether.

The properties of the aether had to be somewhat unusual in order for it to serve the role for which it was invented. It had to fill all space uniformly since the velocity of light is the same throughout space. It also had to penetrate substances such as air, water and glass since light propagates through these materials. Furthermore, this transparent substance could not, in any way, interfere with the motion of corporeal bodies such as the planets since no evidence for the resistance to the motion of corporeal bodies in empty space could be found. In fact, the law of inertia clearly states that the velocity of a body will remain constant al long as no force is acting upon it. The aether can therefore only act on matter through their electric charge. The aether is the medium through which light propagates and through which electric and magnetic forces between charged particles are transmitted.

The experimental detection of the aether became a challenge to the imagination of the nineteenth century scientists, in particular, Albert Michelson, the man who had so accurately measured the speed of light. Michelson argued that since the earth is in motion about the sun it must be moving through the aether. Therefore, a beam of light propagating in the direction of the earth'' motion through the aether would have the velocity c-v relative to the earth. The earth moves through the aether with the velocity v and the light moves through the aether in the same direction with the velocity c and therefore, their relative velocity is c-v. Light propagating in the opposite direction to the earth's motion would have the relative velocity c+v. On the other hand, light travelling in the direction perpendicular to the earth's velocity would have a relative velocity of $\sqrt{c^2 + v^2}$.

Michelson set out to measure the differences in these relative velocities with the help of a physicist named Morley. In 1887, he designed an apparatus, called the Michelson-Morley interferometer, which could detect the differences of these velocities. Using a series of mirrors, light from a single source is divided into two paths and then brought back together again. The light travels along identical paths except for one segment in which the light travels back and forth perpendicular to the earth's velocity on one path and parallel to the earth's velocity on the other path. Because the relative velocity of the light is different along these two paths, as outlined above, it takes longer for the light to travel along the path in which it moves only parallel to the earth's motion. Because of this time delay, one expects to the two beams of light to interfere with each other. The arrangement of light source, mirror and telescopic detector were all mounted on a huge piece of sandstone which was floating in mercury. This enabled the apparatus to be easily rotated so as to align one of its axes with the earth's velocity. After carefully and slowly rotating their interferometer, Michelson and morley were unable to measure any interference.

The inability of Michelson and Morley to observe the effects of the aether was very disturbing to the physicists of their time. Their experiment was repeated and refined a number of times, always with the same result. One of the preoccupations of physicists during this period was an attempt to explain away the results of the Michelson-Morley experiments.

One of the first attempts in this direction was to claim that the aether was dragged along by the earth's motion and hence, the detection of the relative motion of the aether and the earth was impossible. This explanation was easily dismissed, however, since the phenomenon of aberration would not have been detected if the aether was dragged by the earth. One observes a star at an angle $\alpha = \frac{v}{c}$ because of the velocity v of the earth perpendicular to the direction in which the starlight is propagating. If the aether was dragged along with the earth, then, the starlight would also move with the velocity v in the direction of the earth motion and hence, it would no longer be necessary to tilt the telescope through the angle of aberration to capture the starlight.

The Irish physicist, Fitzgerald, proposed that the negative result of the Michelson-Morley experiment could be explained if one assumes that matter in motion contracts in the direction of its motion. He claimed that the contraction, which increases as the velocity of the matter increases, was due to the pressure of the aether wind encountered as the body moved through the static aether. If a body has a length, L_0, at rest then, as a result of its motion, it would have the length $L_0\sqrt{1-v^2/c^2}$. Unfortunately, this contraction could not be measured since any meter stick employed to measure this contraction would also contract. This reduces Fitzgerald's hypothesis to a rather ad hoc status. His idea is not without merit, however. For one thing, Fitzgerald, using his formula, correctly predicted that no material body could ever travel faster than the speed of light, c, since at this speed the length of the body,

$$L = L_0\sqrt{1-v^2/c^2}$$

goes to zero. Secondly, the contraction which he attributed to an actual physical force reappeared later in Einstein's Theory of Relativity with a more sophisticated interpretation. Finally, Fitzgerald's idea stimulated the Dutch physicist, Lorentz.

Lorentz reasoned that the mass of a charged particle would increase as its length decreased. Assuming that the mass of a particle was due to the potential energy of its own charge, he argued that as the particle was crowded into a smaller space, its potential energy would increase and hence, its mass would increase. Since the potential energy is inversely proportional to the distance, he predicted the product of the mass and the length of the particle would remain constant and hence, the mass would increase according to the formulae $M = \frac{M}{\sqrt{1-v/c^2}}$ where M_o is the mass of the particle at rest and v is its velocity relative to the aether. This increase in mass was actually detected in 1900 by W. Kauffman. This result would also be incorporated in Einstein's Theory of Relativity five years later.

Although the Lorentz-Fitzgerald interpretation of the Michelson-Morley experiment contained elements of Einstein's future theory of relativity, they represented a rear-guard action to preserve the notion of the aether. Einstein, unburdened by the traditions of the past, regarded the negative result of the Michelson-Morley experiment as evidence that the luniniferous aether of Maxwell simply did not exist. It is likely that he was aided in arriving at this conclusion in 1905 by the results of his work on photo-electric effect in the same year. In 1900, Planck had discovered that, apparently, the energy of light was quantized in discreet bundles of energy called photons. Einstein's work on the photo-electric effect in 1905 collaborated this view and showed that light in the form of photons sometimes behaves like a particle.

For the present, the significance of Einstein's view that light can also behave like a particle is that the concept of an aether is no longer necessary for understanding the propagation of light through a vacuum. Light can propagate in the form of particles, and hence, the whole notion of an aether can be jettisoned. This is exactly what Einstein did. He went further. He also concluded that the light that traveled along the two different paths of the Michelson-Morley interferometer arrived at the same time because the velocity of light remained constant over the two paths and did not change as a result of the earth's velocity. This deceptively simple interpretation of the Michelson-Morley result forms the foundation of Einstein's special theory of relativity: the velocity of light in free space is the same in all directions of space and is independent of the motion of the light source or observer. This extremely revolutionary idea violates our intuitive notions of space, time and motion.

Let us first consider the addition of velocities. If I were to throw a ball from a moving car in the same direction as the motion of the car, then we all know from our own experience that the velocity of the ball is greater than the same ball thrown with the same force from a stationary car. It is common knowledge that the velocity of the ball and the car add. Is the same thing true of the velocity of light emitted from the headlights of our car?

If we were to carry over our experience from the ball thrown from the stationary car and the moving car, we would surely think that the light from the headlights of the moving car would have a greater velocity than the light from the headlights of the stationary car. This conclusion, which we have reached intuitively, can also be supported logically by arguing in the following manner: If we must add the velocity of the car to the velocity of the ball, then we should add the velocity of the car to the velocity of light. The conclusion that the velocities of the car and light should add, nevertheless, contradicts the results of the Michelson-Morley experiment. Although this contradicts both our logic and intuition, it teaches us that these velocities simply do not add.

The reason that it was so difficult for Fitzgerald, Lorentz, and their contemporaries to arrive at Einstein's simple, literal interpretation of the Michelson-Morley experiment was that it violated their intuition. For them, Einstein's conclusions weren't natural. We experience the same discomfort in understanding and accepting Einstein's Theory of Relativity for it violates our intuition as well. His theory is based on the apparent paradox that the velocity of the car plus the velocity of light is equal to the velocity of light.

But if Einstein's Theory of Relativity violates our intuitive notions of space, time and motion, why should we accept his interpretation rather than the Lorentz-Fitzgerald one? The reason is simply that the Lorentz-Fitzgerald interpretation represents a dead end. It is an ad hoc explanation of the Michelson-Morley experiment which, aside from the increase of mass, did not make any other predictions which could be tested experimentally. Einstein's interpretation led to a theory which made a number of measurable predictions. All of the experimental tests that have so far been performed have all corroborated Einstein's Theory of Relativity.

Since the purposes of a scientific theory are to explain empirical observations rather than reinforce our intuitions, the choice of interpretation is obvious. We must learn to live with what seems to us an apparent paradox. If we consider for a moment, however, that our intuition concerning the addition of velocities developed only through the consideration of velocities, much less than the velocity of light, perhaps we may resolve this paradox. After all, the fastest velocities we ever consider in our day-to-day existence are never more than the speed of sound which is only 3 × 104 cm./sec. Or approximately one-millionth the velocity of light. Perhaps we may grant now that it is possible that the intuition we developed with super-slow velocities, like the speed of sound, do not apply when we are dealing with the velocity of light.

While this reasoning does not completely resolve the paradox in our minds, it helps us to understand how it is the physics with velocities of the order of c can be so different from the low velocity physics that we intuitively identify with because of our experience. Bearing this in mind, perhaps we will not feel so uncomfortable with the ideas of relativity. Some discomfort is inevitable, however. We will notice that certain relativistic effects such as the Fitzgerald contraction encountered earlier disappear as the velocity involved becomes small compared with the velocity of light. For example, a mass of a particle travelling at the speed of sound increases only one part in 10^{12}.

THE SPECIAL THEORY OF RELATIVITY

The special theory of relativity is based upon the assumption that the velocity of light is independent of the motion of both the observer and the source of light. With this assumption, Einstein destroyed the notion of absolute motion. Before he formulated his theory of relativity, physicists believed that there existed a frame of reference absolutely at rest and that the motion of all objects in the universe was to be taken with respect to this absolutely stationary frame of reference. It was in this frame of reference that the aether sat motionless and, as a consequence, in this frame and only this frame, the velocity of light was exactly equal to c.

If an observer was in a frame of reference and he wanted to determine the absolute motion of his frame of reference, all he had to do was to measure the velocity of light in his frame of reference. If the velocity of the light in his frame was exactly equal to c, then his frame was absolutely at rest. From the deviations of this velocity from c, he could determine the absolute motion of his frame of reference with respect to the stationary aether. This was the aim of the Michelson-Morley experiment which as we know filed to detect any motion of the earth with respect to the aether.

The negative result of the Michelson-Morley experiment led to Einstein's formulation of the Theory of Relativity in which the velocity of light is the same in all frames of reference independent of their motion. It is, therefore, impossible to determine the absolute motion of a frame of reference since the velocity of light will always be the same. All motion is relative. No frame of reference is preferred over any other. Einstein formulated this concept in terms of his principle of relativity:

The laws of nature are identical in all uniformly moving frames.

By a uniformly moving frame, we simply mean a frame of reference which is not undergoing acceleration. In other words, the laws of nature described with respect to the stationary earth or with respect to the interior of a train moving at a constant velocity with respect to the earth are the same. (We are ignoring, for the purposes of this discussion, the motion of the earth in its orbit about the sun or its rotation). If the train was accelerating or deaccelerating, the laws of physics in the frame of the train would not be the same as those on the earth. The reason for this is a mass with no force acting upon it would experience a fictitious force in the accelerating train but not on the stationary earth. The fictitious force arises because the mass will continue to move at a constant velocity due to its inertia independent of the motion of the train. If the train accelerates then the constant motion of the mass will no longer be constant with respect to the accelerating train.

With respect to the train, the mass will appear to be accelerated by a fictitious force. This is a familiar experience to all who have been thrown forward in a moving train or bus which suddenly deaccelerates. When describing the laws of physics it is, therefore, wisest to remain in a non-accelerating frame of reference so as to avoid the presence of friction forces.

In our discussion, therefore, we will automatically limit ourselves to frames of references for which there are no fictitious forces. Therefore, unless otherwise mentioned, the reader may assume that the frame of reference under discussion is one undergoing uniform motion. Einstein's principle of relativity states that all such frames are equivalent and the laws of physics described in each of these frames are identical.

The idea that all motion is relative was as devastating to Einstein's contemporaries as Copernicus' notion that the earth was no longer the centre of the universe. It took people over one hundred years to accept the idea that the earth actually moved. Once this idea was accepted, thinkers became used to the idea that the earth was moving with respect to some frame of reference absolutely at rest.

So, although the earth was not at rest, at least there was some place in the universe at rest, a place where one could anchor one's thoughts. The concept of absolute motion and absolute space which evolved essentially from the physics of Copernicus, Galileo, and Newton was elevated to the heights of an *a priori* truth by the philosoper, Immanuel Kant. This is an indicator of how engrained the concept of absolute motion became in the minds of Western thinkers.

Kant could not conceive of the possibility that space could be structured in any other way. Einstein's principle of relativity based on experimental fact, completely destroyed the validity of this so-called *a priori* or absolute truth. Instead of there being one frame of reference at rest, Einstein showed that any frame of reference not accelerating may be considered at rest and the motion of all the other objects in the Universe may be taken with respect to it.

We have all experienced the relativity of motion while riding in a train or a subway car. Remember the sensation when your car was at rest in the station and you looked out the window to see another car besides yours moving out of the station. For a moment if feels as though your car is in motion. In fact, unless you can observe the ground or some other fixed structure in the station it is impossible by just looking at the other car to determine whether you car or the other car is in motion. This example illustrates the relativity of motion. Before Einstein, physicists believed they could always tell which car was really in motion. Einstein has shown this is pure folly. You can determine which car is moving with respect to the earth but there is no way of determining whether the earth and the stationary car are really at rest and the moving car is in absolute motion or vice-versa.

In the Einsteinian universe, the whole notion of space changes; it is no longer defined in the absolute sense in which it is in the Newtonian universe. For Einstein, space does not exist per se. Space is a relationship between physical objects. If there were no objects in the universe, there would be no space in Einstein's universe, whereas in Newton's universe, there still would be space. Space has an *a priori* existence for Newton. It is the container which holds the universe. If one considers a finite universe from the Newtonian point of view one can discuss the space outside the universe. In the Einsteinian world view, one can not. Space for Einstein is a relationship, whereas space for Newton is something which has existence and reality.

Let us illustrate there two contrasting viewpoints by considering the space between my two outstretched hands which we hold up before you. What happens to that space when we put my hands behind my back? From the Newtonian point of view, the space is still there, just sitting there ready to be filled, if necessary. From the Einsteinian point of view, once my hands have disappeared, we can no longer talk about the space between them. The reasoning is similar to the one employed in the riddle: Where does my lap go when I stand up? My lap is the relationship of my legs when I am seated and no longer exists when I stand. In a similar fashion, according to Einstein's way of thinking, space is also a relationship of physical objects which disappears when those objects disappear.

The resistance to the adoption of Einstein's ideas was due to the radical change required in the way one regarded space. This, plus the non-addition of the velocity of light, made his theory very unappealing to his contemporaries. The Einsteinian world view was also unpopular because of the way in which time was regarded. In the Newtonian world view, time had an absolute status, like space. In the minds of scientists and philosophers, there was a clock somewhere in the universe which went on ticking through eternity by which all events took place. In fact, one of the popular images of the universe was that it was a giant clock which kept an eternal time. The absolute nature of time, like space, was also given the status of an *a priori* truth in Kant's philosophical system.

In the Einsteinian world view, on the other hand, time is merely the relationship between events. It does not exist per se. If there were no events in the universe, then there would be no time. We have all, no doubt, experienced this concept psychologically. When nothing is happening to us, we lose our sense of time. Men who have spent long periods of time underground with nothing to do lose their sense of time altogether. We are most aware of time when we have a great number of things we must do.

Let us consider a universe with a finite lifetime so that there are two events corresponding to the birth and death of the universe. Within the framework of Einstein's world view, one can not conceive of or discuss the time before the creation of the universe of the time after its destruction. These times simply do not exist as they do in the Newtonian world view. As with space, time for Einstein is a relationship, whereas, for Newton, time is something which exists and is real. Time, in the Newtonian sense, is that which a clock tells. Just as a clock, if properly wound, would continue to run, so time for the Newtonian continues to pass.

In order to develop a feeling for time in Einstein's world, we should examine how the concept of time developed before the existence of the clock. In essence, our sense of time developed from our ability to order or sequence events. If I were to whistle, stamp my feet and then clap my hands before a group of students, there would be a general agreement as to the order in which I performed each of the events. There might be disagreements as to the duration between events, but there would be universal accord concerning the sequencing of the events.

This ability for us to sequence events provides the foundation for our sense of time. As man began to associate the events in his life with the motion of astronomical bodies such as the sun and the moon, his concept of time became more sophisticated. He began to describe the intervals between events in terms of the number of times the sun had risen between the two events or in terms of the number of moons. Later, he divided the interval between successive sunrises into smaller units called hours. Each hour corresponded to a different position of the sun in the sky. He began describing the intervals between events in terms of the change of the sun's position in the sky.

Finally, instead of always relating events to the motion of the sun or moon, man created an instrument called the clock which, like the sun in the sky, executed a regular type of motion.

A correspondence was then made between events in the real life and events on the clock. So, for example, when we say that we arose at 8 o'clock in the morning, where we are in essence saying is that the event of our arising and the event of the hands of the clock indicating 8 o'clock coincided.

If we think of a clock as an instrument which performs a regular and repetitive motion to which we may refer other events, we will be more in the spirit of the Einstein view of time. If we consider the clock as an instrument which measures the passage of time, we shall retain the Newtonian concept of time.

The Newtonian concept of absolute motion is destroyed in the theory of relativity because the velocity of light is a constant, in all frames of reference. As a consequence, all motion is relative and space is no longer absolute as it was in Newtonian physics.

It is also the constancy of the speed of light which destroys another concept central to Newtonian physics, namely, the absoluteness of time. According to Newtonian thinking, there is only one absolute time which is valid for all frames of reference. In the theory of relativity, on the other hand, time is also relative to the particular frame of reference under consideration. In other words, the time in two different frames of reference is not the same.

In order to illustrate how this follows from the constancy of light, we will consider a hypothetical situation originally suggested by Einstein to explain his ideas. In essence, we will conduct a thought experiment, that is, create an experimental situation in thought.

Let us consider two observers, one sitting on a train moving at a constant velocity with respect to our second observer sitting at rest along the track on which the train is travelling. Two flashes of lightning strike the front and the rear of the train.

The train has a length L and velocity v. The observer sitting precisely in the middle of the train sees the two flashes of lightning hit the train at exactly the same time. He is sitting exactly in the middle of the train and, therefore, the times for the transit of the signals of the lightning hitting the two ends of the train to his eyes are the same. He concludes that the two flashes of lightning struck the train at exactly the same time.

An observer is stationed at rest beside the railroad track so that he is exactly opposite the moving observer precisely at the moment when the light from the lightning striking the ends of the train enters the eyes of the moving observer.

Will the stationary observer also report that the two bolts of lightning struck the train at the same time? In order to determine this, let us consider how he will perceive the two events under consideration. When the lightning actually struck the train, the train was farther down the tracks, at this moment, the stationary observer was actually closer to the front of the train than to the end of the train.

Because of the fact that the velocity of light is a constant, the light from the two bolts of lightning striking the front and the rear of the train travels towards our stationary observer at the same speed, c, even though the front of the train is moving away from him and the back of the train is moving towards him.

Because of the constancy of the velocity of light and the fact that the stationary observer is closer to the front of the train when the lightning strikes, he will see the lightning strike the front of the train before it hits the rear of the train.

He will conclude that the lightning struck the front of the train before it hit the rear of the train. The stationary observer has a completely different sense of time from that of the moving observer. The two events which the moving observer reported to be simultaneous occur at different times in the frame of reference of the stationary observer.

The discrepancy in the timing of the two events by our two observers is due to the fact that the velocity of light is a constant in both frames of reference. If the velocity of light added with the velocity of the train, as our intuition would have us believe or as Newtonian physics would instruct us, then the velocity of the signals arriving from the front and the rear of the train to the stationary observer would not have been the same. The velocity of the light from the lightning striking the rear of the train would have traveled towards the stationary observer with the velocity c - v while the signal from the front of the train would have travelled with the velocity c - v.

The distance from the rear of the train to the stationary observer when the lightning strikes is $\frac{L}{2}\left(1+\frac{v}{c}\right)$ whereas the distance to the front of the train is $\frac{L}{2}\left(1+\frac{v}{c}\right)$. This means that if the velocities had added in the Newtonian sense, they would have arrived at exactly the same time. We must, therefore, attribute the different sense of time possessed by our two observers to the constancy of the velocity of light.

The result of the constancy of the velocity of light will also affect the spatial perceptions of the two observers moving with a constant velocity with respect to each other. The difference in the spatial perception of our two observers manifests itself in terms of the Lorentz-Fitzgerald contraction. Let us consider a stick, which, at rest, has the length L_o, moving with respect to a stationary observer with the velocity v parallel to its length. The stationary observer will discover that, in his frame of reference, the stick has suffered a contraction and only has the length $L_0\sqrt{1-v^2/c^2}$. An observer moving with the velocity, v, of the stick will discover, however, that in his frame of reference, the stick, at rest, still has the length L_o. In other words, as a result of its motion, an object appears to a stationary observer to have contracted by the Fitzgerald factor, $\sqrt{1-v^2/c^2}$. (Note that the factor $\sqrt{1-v^2/c^2}$ has a value less than one but greater than zero.)

The contraction of the object only takes place in the direction parallel to its motion. The length of the object perpendicular to its motion will appear the same to the stationary observer. A moving square will appear, to the stationary observer, to shrink into a rectangle whose shorter side is parallel to its motion, and a moving circle will become an ellipse with its major axis perpendicular to its motion.

We have plotted in the length of a meter stick, as it appears to a stationary observer, versus its velocity. Naturally, when its velocity is zero, its length still appears to be one meter. When the velocity of the stick is 5c then its length, to the stationary observer, seems to be 88 meters, as its velocity increases to 88c, its length appears now to be only one-half of its original length.

When its velocity reaches 999c, then its length will appear to be only 045 meters. As the velocity of the stick approaches closer and closer to the velocity of light, its length appears shorter and shorter to the stationary observer. An object travelling at the velocity of light would literally disappear from view. This is impossible, however, since no particles with mass can travel faster than the velocity of light.

The interpretation of the Lorentz-Fitzgerald contraction within the framework of the theory of relativity is quite different from the one originally made by Lorentz and Fitzgerald. According to these early workers, the moving meter stick actually undergoes a physical contraction due to pressure of the aether wind arising from the absolute motion of the meter stick with respect to the aether. The contraction is due to the absolute motion of the meter stick.

From their point of view, if the observer were to move and the meter stick remained stationary, the observed would not observe a contraction. In the theory of relativity, on the other hand, the contraction is observed whether the stick moves with respect to the observer or the observer moves with respect to the stick. Furthermore, the stick does not actually physically contract. An observer moving with the stick does not observe a change in its length. The stick appears shorter to the stationary observer because his concept of space is different than that of the moving observer. They perceive the same physical phenomenon differently.

Not only are the spatial perceptions of the stationary and moving observers different but their temporal perceptions differ also. We saw evidence of this earlier when we considered the example of the lightning striking the train.

The two observers had a different notion of the simultaneity of the lightning flashes. Perhaps even more surprising is the fact that a stationary observer will observe that time in a moving frame of reference actually slows down. A moving clock runs slower than a stationary clock.

For example, if a clock is moving at the velocity of 866c with respect to a stationary observer, the time required for the passage of one hour on the moving clock will take two hours on a clock at rest in the stationary frame. In general, a time interval that takes to seconds in the moving frame of reference will take

$$t = \frac{t_0}{\sqrt{1 - v^2/c^2}}$$

seconds in the stationary frame.

The time elapsed in the stationary frame for the passage of one hour of a moving clock is plotted versus the velocity of the clock. Notice that, as the velocity of the clock approaches c, the time for one hour to elapse in the moving frame, requires an infinite amount of time or, in other words, time comes to a standstill in the moving frame.

The slowing down of time in a moving frame of reference must seem like science fiction to the average reader who has had no experience with fast-moving objects. To an elementary particle physicist who deals all the time with sub-atomic particles which travel at velocities near the velocity of light, the slowing down of time in a moving frame of reference is a very real thing.

It, in fact, helps him in his study of certain short-lived elementary particles called mesons. Mesons are created with a high energy accelerator as the result of the collision of protons with other protons or nuclei. One of the mesons created in such a collision is the short-lived pi meson which lives, on the average, about 10^{-8} sec before decaying into other elementary particles. The average lifetime of $10^{-8}/\sqrt{1 - v^2/c^2}$ sec quoted for the pi meson is the value of the meson's lifetime when it is observed at rest. As the velocity of the meson, with respect to the observer in the laboratory increases, so does the lifetime of the meson.

The reason for this is that time for the moving meson, like a moving clock, slows down and hence, the lifetime of the meson at rest, 10^{-8} sec, takes longer to pass in the laboratory frame. In fact, the measured lifetime of the meson, as a function of its velocity, is just T = 10^{-8}/sec, exactly the value one would predict on the basis of the theory of relativity. The extra time that the meson lives in the laboratory frame, as a result of its velocity, has a practical aspect.

It gives the elementary particle physicists enough time to study the meson before it decays. If it was not for the time dilation of the meson's lifetime, the meson would decay after 10^{-8} sec and there would not always be enough time to study it in the laboratory frame.

The slowing down of time in a moving frame of reference is not just a theoretical idea but something which occurs daily in the laboratory of the elementary particle physicists. Although the time in the moving frame appears to slow down for the stationary observer, an observer in the moving frame does not experience time slowing down.

As far as he is concerned, his clocks run at the same rate. An hour is an hour and the lifetime of a pi meson is only 10^{-8} sec. This is to be expected since, from his point of view, he is at rest and hence, expects time to flow naturally. In fact, from his point of view, the stationary frame of reference is in motion with respect to him, and therefore, he observes time slowing down in this stationary frame.

As paradoxical as this may seem, both the stationary and the moving observer will observe each other's clock slowing down. Each believes that the time is dilated in the others' frame. How is this possible? How can time slow down for both? In addition to this mutual time dilation, each observer will also observe the others' meter stick to be shorter. Is this possible?

Which meter stick is really shorter? Which clock is really telling the correct time? We cannot ask questions like this, because in asking such a question, we are assuming the existence of the absolute space and absolute time which characterized Newtonian physics. Within the framework of Einsteinian physics, we cannot discuss such things as the real length of the stick or the real time of the clock.

The observations of one observer are as real or as valid to him as those of another observer are to him. We simply must accept the notion that two observers moving with respect to each other do not empirically observe the same things. The unstable meson only lives $\frac{10^{-8}}{\sqrt{1-v^2/c^2}}$ sec in the frame in which it is at rest and it really lives sec in the frame in which its velocity is v. The quantities that an observer measures are only relative to his particular frame of reference. One can not make an absolute comparison. Both observers will really observe the others clock as slowing down.

The phenomenon of relativistic time dilation might, some day, make interstellar space travel possible. The closest star, Alpha Centauri, is four light years away. A light year is a unit of length equal to the distance that a beam of light travels in one year or about six trillion miles. This means that, since the velocity of light is the fastest speed attainable no matter how proficient our technology became, it would take a minimum of eight years to send an astronaut to Alpha Centauri and back.

Since there are only a few stars in this corner of our Milky Way galaxy that are within fifty light years, the minimum time to send a space craft to most stars would be 100 years or more which is greater than the lifetime of even our healthiest astronauts. Even if we were to develop a space craft which could travel at velocities near the speed of light, it does not seem that it would be possible to send a man to a distant star and have him live long enough to return alive.

Because of the relativistic time dilation, however, this is actually possible. The reason for this is that, while the astronaut is travelling to a distant star at a velocity close to c, his clock would slow down with respect to an earth clock. So, for example, if he was travelling at .999c to a star 99.9 light years away, he would require 100 earth years to make the journey to the star. In his own frame of reference, however, he would age only 4.47 years. (These two times are related by the formulae $t = \dfrac{t_0}{\sqrt{1 - v^2/c^2}}$, where t = 100 years, t_o = 4.47 years and v = .999c).

One might wonder how it is possible for the astronaut to travel 99.9 light years in only 4.47 years. The reason is, that in his frame of reference, the distance to the star is contracted by exactly the same factor that his time is dilated with respect to the earth.

In other words, in his frame of reference, he sees a star the distance

$L999\sqrt{1 - \dfrac{(.999c)^2}{c^2}} = 4.46$ light years away travelling towards him with the velocity .999c. Therefore, it takes only 4.47 years for the journey to the distant star.

The astronaut will only require nine years, according to his clocks, in order to make the round trip to the star and back again. However, when he returns to earth, he will discover that, during the nine years he was away, 100 years had elapsed on earth. The astronaut will have aged only nine years but all of his friends will have aged 100 years and hence, would probably all be dead. The astronaut will have experienced not only a round trip to a star 100 light years away, but he will also have experienced travelling 91 years into the future.

The exploits of our hypothetical space traveller have given rise to a famous paradox known as the twin paradox. Let us assume that our space traveller has a twin brother who remains on earth during his flight. The twin on earth will expect his brother in space to age at a slower rate as a result of his motion, as we explained above.

The formulators of the twin paradox maintain, however, that the space traveller will also expect his brother on earth to age at a slower rate. According to their argument, during his flight the astronaut, in his frame of reference, will see his brother on earth moving away from him and then towards him. As a consequence, he will observe his brother's clock slowing down and hence, he, too, will expect his brother on earth to age at a slower rate than himself.

One is faced with a paradox. Which brother will actually be the younger when the two brothers are reunited back on earth? The brother who actually traveled to the star and came back to earth will be the younger. In resolving the paradox, we must take into consideration the effect of the traveler's motion on his observations.

The paradox is only an apparent one because it is based on an assumption which is not valid. When we considered the astronaut's observations of the earth-bound clocks, we assumed that his observations were made in a frame of reference in uniform motion. This assumption is not valid because, in order for the astronaut to return to earth, he must change directions in the middle of his trip. This naturally involves acceleration or non-uniform motion.

The astronaut was no longer in an inertial frame of reference during the time that he is turning around and as a result, his observations of time on the earth are affected. It is true that, while he is travelling at a constant velocity, he observes the earth clocks slowing down with respect to his. During the time that he is turning around, however, he observes the earth clocks as speeding up. Although the time, in his frame to reverse directions might not take very long, it is during this period that he observes the passage on earth of 91 years that will separate the earthbound clocks from his clock when he returns to earth.

Now that we have experimental proof that a space traveller will come back younger than his twin, we know how to build a time machine. H.G. Wells, in his famous short novel, The Time Machine, describes a device that allows its passenger to travel into the future. When this story was first written, the idea of a time machine was pure fantasy. This idea is nno longer a fantasy. If it were possible to build a space craft that allowed a man to travel at velocities near the velocity of light, then one could travel into the future by making a round trip journey.

The closer to c at which one travels, the farther into the future one would be projected. For example, if one wished to devote one year of his life to travelling into the future and one's space ship would travel at the velocity v, then he would find himself projected $\frac{1}{\sqrt{1-v^2/c^2}}$ years into the future.

If he traveled at v = .99995c he would be projected 100 years into the future whereas, if he traveled at v = .999999995c, he would be projected 10,000 years into the future. If one was unhappy about the point in time one arrived at, one could always get back into the space ship and travel a bit further.

The only problem with our time machine, however, is that once one traveled into the future, it would not be possible to travel back to the time one started from. H.G. Wells' time machine did not suffer from this problem. His device was purely a product of his imagination. Our device, while not practical at the moment, does not violate any of the laws of physics, however. Science fiction writers, since the advent of the theory of relativity, have exploited the relativistic time machine, described above, to project their characters into the future. In order to project backwards in time, these writers usually have their character travel faster than the speed of light.

But how does travelling faster than the speed of light project one backwards inn time? It really cannot. If one could travel faster than the speed of light, one could overtake light signals that were emitted from earth a long time ago in the past. By intercepting these signals, one could observe events that occurred hundreds of years ago such as the building of the pyramids or the extinction of the dinosaurs.

While this is not equivalent to actually living in the past, it represents travelling into the past in a certain sense. One can see how a science fiction writer might be able to imagine his character travelling backwards in time. Unfortunately, it is not possible for us to travel faster than c, so there is no hope for seeing back into the past. We can, however, anticipate travelling into the future but this will require tremendous advances in space travel technology. At the moment.

Our claim that man cannot travel faster than the speed of light is based on experimental facts. Accelerators have been built which accelerate electrons and protons through electric potentials. The velocity of the charged particles increases as a result of their absorbing electrical energy. It is found that, as the velocity of the charged particle approaches c, a large amount of energy is required to increase the particles' velocity by even a small amount. One finds, in fact, that no matter how much energy is transferred to the particle, its velocity never reaches c. Apparently, the velocity of light is the ultimate velocity of the universe. Since all matter is composed of particles, such as electrons and protons, these experimental results show that the velocity of light may never be exceeded.

In order to understand why particles cannot go faster than c, it is necessary to examine the behaviour of the mass of a particle as a function of its velocity.

Inn fact, the mass of the particle becomes infinite as the velocity of the particle approaches c. Since the mass of a particle depends on its velocity, it is useful to define its rest mass, which is the value of its mass when it is at rest and which we denote by m_0. The mass of a moving particle, with respect to a given frame of reference, is related to its rest mass, m_0 and its velocity v, with respect to that frame by $m = \frac{m_0}{\sqrt{1 - v^2 / c^2}}$ this relationship.

It is the increase of mass with velocity which explains why the ultimate or highest velocity of a particle is c. The mass of a particle is a measure of its resistance to acceleration by a force. As one accelerates a particle to velocities approaching c, its mass increases significantly making it more difficult to accelerate. Because the mass becomes infinite, exactly in the limit that the velocity becomes c, it is impossible to generate an infinite force.

This explains why the ultimate velocity of matter is c. No massive particle actually obtains the velocity c since this would require an infinite amount of energy. Particles whose rest mass is zero, however, such as the proton and the neutrino, travel at the velocity c. In fact, they can never have a velocity less than c.

The expression for the energy in the theory of relativity differs quite markedly from its expression in Newtonian physics. Einstein found, however, that there is a simple relationship between the energy and the mass of a particle, namely, the energy of a particle is equal to its mass times the velocity of light squared, $E = mc^2$.

If we express the relavistic mass in terms of the particle's rest mass, m_0, and velocity, v, then the familiar expression for energy, $E = mc^2$ becomes $E = \frac{m_0 c^2}{\sqrt{1 - v^2 / c^2}}$ Einstein's expression for the energy for velocities much less than the speed of light, reduces to $E = m_0c^2 + 1/2m_0v^2$, which differs from Newton's no-relativistic expression for the energy, $1/2m_0v^2$

by the addition of the extra term, m_0c^2. This extra term is the rest mass energy of the particle and is equal to the energy of the particle at rest. According to the interpretation of this term given by Einstein, the particle has energy simply by virtue of its rest mass.

This means that mass and energy are equivalent. If this is true, then one should be able to convert mass into energy and energy into mass just as one converts energy into heat and heat into energy. This was one of the major predictions to grow out of Einstein's theory of relativity.

It was on the basis of this prediction that work on the construction of the atom bomb began in the U.S.A. during World War II. It is sad that the prediction of a man who struggled for world peace, as Einstein did, should be verified in a device of war and destruction such as the atom bomb. The atom bomb was indeed the first device to convert mass into energy on a large scale. Unclear fission had been observed in a laboratory situation earlier.

The energy released in nuclear fission results from the destruction of matter. The total mass of all of the products of the fission reaction are less than the masses of the original uranium nucleus and the neutron which triggers the reaction. Another example of a nuclear reaction in which energy is released is thermo-nuclear fusion, the process responsible for the energy of the sun. The mass of the final product of fusion, the helium nucleus, weighs less than the hydrogen nuclei which combined to form the helium nucleus. Another process whereby energy is converted into matter and back into energy is the process of pair creation. In this process, energy, in the form of light, is converted into the masses of an electron and an anti-electron.

An anti-electron or positron is a particle with the same mass and spin of an electron but which has the opposite charge. An anti-electron, like an electron, is stable by itself.

However, when an electron and anti-electron make contact, they annihilate each other and change back into light energy. The creation and annihilation of electron - anti-electron pairs are processes whereby light energy is converted into matter and then back into light energy. We will study this process in greater detail when we come to our study of elementary particles at which time we shall encounter other examples which demonstrate the equivalence of mass and energy.

Before completing our study of the special theory of relativity, we turn to one of its first applications in which Einstein showed the equivalence of the electric and magnetic forces. Maxwell had shown, through his equations, that the electric ad magnetic fields were intimately connected to each other. Einstein was able to demonstrate the actual equivalence of these forces, however, by using his principle of relativity which states that the law of physics for two observers moving at a constant velocity with respect to each other are the same.

Einstein asked us to consider two charged particles moving along two parallel lines at constant speed with respect to a stationary observer. The stationary observer will observe two forces between the charged particle. One force, the electric force, arises solely by virtue of their charge. The other force, the magnetic force, arises by virtue of both their charge and their velocity.

From the point of view of an observer moving with the same velocity as the charged particles, there will only be an electric force between the charges. He will not observe a magnetic force because, with respect to him, the charges are both at rest. The interaction of the charged particles is identical for the two observers because of the principle of relativity. Hence, the electric and magnetic force of the stationary observer is equivalent to the electric force of the moving observer or, in other words, the electric and magnetic forces are the same. The magnetic force is just an electric force in motion.

THE GENERAL THEORY OF RELATIVITY

Einstein's Special Theory of Relativity deals with the effects of the uniform motion on the relative space and time perceptions of moving and stationary observers. Discussion within the special theory of relativity was always limited to non-accelerated motion. As soon as accelerated motion was encountered, as in the twin paradox, we found ourselves beyond the scope of the special theory of relativity.

It is in Einstein's General Theory of Relativity, published in 1916, that accelerated motion is taken into account. The general theory of relativity generalizes the results of the special theory and deals with the effect of accelerated motion on space and time perception. The general theory also deals with the effects of a gravitational field on space and time perception by showing the equivalence of accelerated motion and the existence of a gravitational field. This is accomplished by exploiting the fact that the inertial and gravitational masses of all particles are equal. Finally, the general theory of relativity provides a theory of gravity which replaces Newton's theory of gravity.

We discovered that, within the framework of the special theory of relativity, it is impossible to determine the absolute motion of a frame of reference through space as long as that frame is not undergoing acceleration. Once a frame of reference undergoes acceleration, however, it is impossible to detect this motion because of the fictitious forces which arise as a consequence of the acceleration.

These forces are fictitious because, in actuality, the acceleration of a body, with respect to the frame of reference that these fictitious forces seem to cause, it only apparent. In actuality, the body continues to move with constant velocity because of inertia but, relative to the accelerating frame, it appears to be accelerating. The fictitious force causes real motion with respect to the accelerating frame, however, and hence, is perceived as a gravitational force.

Einstein exploited this fact and claimed that the effects of the accelerated motion are equivalent to those of a gravitational force. In order to make this assertion, he explained the fact that the gravitational and inertial masses are the same and hence, all particles accelerate at exactly the same rate in a gravitational field. This is also true of the apparent acceleration of the particles in an accelerating frame of the reference. They, too, accelerated at the same rate.

In order to illustrate the equivalence of accelerated motion and the action of a gravitational field, let us consider the two elevator cars in Fig 13:1. One elevator car is sitting at rest on the surface of the earth. The other elevator car is situated in outer space completely removed from the effects of any gravitational field.

The elevator car is being accelerated upwards with a constant acceleration, g, equal to the acceleration of the particle falling near the surface of the earth. We claim that an observer in one of these cars would not be able to determine in which car he was situated because his experiences in both cars would be identical.

We would experience a gravitational pull towards the floor of the elevator in both cars. If he dropped two objects, let us say, an iron ball and a wooden ball, in the elevator car on earth, these two balls would fall at exactly the same rate because of the equality of their gravitational and inertial masses.

If our observer were to drop the balls in the elevator car in outer space, he would experience the two balls accelerating towards the floor at exactly the same rate as they did in the elevator car on earth. The reason for this is as follows: As the observer, releases the two balls from his hand, the balls will no longer be accelerated upwards. They will remain at rest in an inertial frame of reference which was moving at the same speed as the elevator at precisely the time the balls were released.

The elevator continues to be accelerated, however, and hence, the floor of the elevator will accelerate up at the rate, g, towards the balls, until the floor hits the balls. To an observer in the elevator, however, it appears as though the balls have fallen to the floor with the acceleration, g. to an observer in the accelerating elevator, masses behave exactly as they do in the elevator on earth. The acceleration of his elevator car acts exactly like a gravitational field.

If the direction in which the elevator car was being accelerated changed direction so that the elevator was being pulled down with an acceleration, e.g., then there would be an effective gravitational field pulling things towards the ceiling of the elevator instead of towards the floor.

If such an elevator was placed near the surface of the earth, then the effects of the gravitational field produced by its acceleration down would exactly cancel the gravitational field of the earth. This is exactly what happens in an aircraft which is experiencing free fall.

The effective gravitational field created as a result of its being accelerated down by the earth, exactly cancels the gravitational field of the earth and hence, observers in the free falling craft experience no gravity. Objects will float around inside the aircraft just as they do when an astronaut travels to the moon and is no longer within the influence of the earth's gravitational field.

Following Einstein, we have established that the laws of physics describing the motion of massive particles are the same in the elevator car sitting in the earth's gravitational field at rest and in the elevator car being constantly accelerated upwards with an acceleration, g. Einstein concluded that, on the basis of this demonstration, all the laws of physics would be identical in these two frames of reference. He expressed his hypothesis in terms of the equivalence principle which states the following: the phenomena in an accelerating frame of reference are identical with those in a gravitational field.

The equivalence principle forms the heart of the general theory of relativity. Einstein exploited this principle to successfully predict that, a beam of light, i.e., a beam of photons, would be bent by a gravitational field. This is somewhat surprising in view of the fact that the rest mass of a photon is zero.

Although the rest mass is zero, the photo still has energy, and since Einstein showed that mass and energy are equivalent, perhaps light can also be affected by gravity. To demonstrate this, we turn to the propagation of light in an accelerating elevator car.

Let us consider a beam of light propagating perpendicular to the direction of the acceleration and entering the elevator from one wall and exiting the elevator on the opposite wall. The beam of light will appear bent to an observer inn the elevator.

The reason for this is that, by the time the light beam has propagated from one wall to the other, the elevator has moved upwards because of its acceleration. The beam of light, however, is unaffected by the elevator's acceleration ad hence, continues to propogate along the same straight line it was moving along before it entered the elevator.

Relative to the accelerating elevator, however it appears to exit at a point below the one it entered.

The beam of light will appear, to a observer in the accelerating elevator, to have been bet by the same gravitational field that causes the objects he drops to also fall to the floor. If, instead of a beam of photons, we had sent a beam of massive particles through the elevator, they would behave exactly like the beam of light.

An observer in the accelerating frame will conclude that the paths of both the massive particles and the light were bent because both the massive particles and the light were attracted by the gravity. At this point, Einstein exploits the equivalence principle.

He argues that, since the laws of physics are the same in the accelerating frame and the stationary frame sitting in a gravitational field, then, one can expect the beam of light to be bent by a bonafide gravitational field such as that generated by the sun.

On the basis of this argument, he predicted that starlight passing close to the sun would be bent by its gravitational field and hence, during a solar eclipse, the position of a star close to the sun would be displaced from its normal position.

The first opportunity to measure the effect of the gravitational pull of the sun on starlight came during the total solar eclipse of 1919. The expedition of British scientists who traveled to Africa to observe the eclipse, found that the position of stars near the sun had indeed changed, as Einstein had predicted. This measurement verified the validity of the equivalence principle.

Einstein further exploited the equivalence principle to determine the effects of a gravitational field on the space and time perceptions of an observer. Let us consider from the point of view of a stationary observer in an inertial frame, a clock in an accelerating elevator car.

By virtue of the fact that the clock develops a velocity with respect to the stationary observer, as a result of the elevator's acceleration, we expect the clock to slow down. Hence, the stationary observer will observe a time dilation in the accelerating frame of reference. By virtue of the principle of equivalence, we would also expect that a clock sitting in a gravitational field would also slow down.

This gravitational time dilation has been experimentally verified in two separate experiments, both of which involve 'atomic clocks'. An atom is like a clock. The electrons orbit the nucleus of the atom with a fixed periodicity. The frequency of light that an atom emits is related to the periodicity of the electron's orbits. Since two identical atoms will radiate the same discreet frequencies or spectral lines if they are in the same inertial frame, a change in the frequency of the radiation of an atom can indicate a change of the time in the frame of reference in which it is situated.

The first detection of gravitational time dilation was made by observing the slowing down of the 'atomic clocks' on the surface of the extremely dense white-dwarf stars. The expected shift of the spectral lines towards the red was observed. A more precise observation of the gravitational time dilations was made by Pound in a laboratory setting at Harvard University. Pound compared the spectral lines of two atomic clocks separated by a mere 70 feet in the earth's gravitational field. He found that, the 'atomic clock' sitting closer to the earth and hence, more strongly influenced by its gravitational field, slowed down more than its counterpart 70 feet above it.

Einstein's General Theory of Relativity describes the effect on space and time measurements of an observer, either imbedded in a gravitational field or else undergoing non-uniform motion.

The General Theory of Relativity also provides a theory of gravitation which competes with Newton's theory of gravity. In addition to providing relativistic corrections to Newton's theory, Einstein's theory attempts to explain how the gravitational interaction arises. Newton's principle of action at a distance is replaced by a field concept. According to Einstein, matter warps the space. The warped space, in turn, affects the motion of the matter contained in it. The sun, for example, warps the space in which the solar system is embedded, creating grooves in which the planets move. Determining the gravitational interaction of matter becomes a matter of geometry.

Before turning to the details of Einstein's model of gravity, we must first discuss the four-dimensional space-time continuum of relativity which replaces Newton's concept of three-dimensional space and absolute time. In the special theory of relativity, we discover an intimate relationship between space and time.

Time is no longer independent of the position or motion of an observer. Shortly after Einstein's special theory appeared in 1905, Minkowski proposed that three-dimensional continuum with time playing the role of the fourth dimension. In this space, every event is described by four co-ordinates or number. Three of these co-ordinates, x, y, and z, describing the spatial location of the event while the fourth component, t, describes the time when the event occurs. These co-ordinates are defined with respect to an inertial frame of reference.

The details of these equations are not important for the purposes of our discussion so don't let their apparent complication disturb you if you are not mathematically inclined. In Newtonian physics, the time in the moving frame would be identical to the time in the stationary frame.

We see, in relativistic physics, that space and time are interwoven. The contraction of length and the dilation of time that the observer in the stationary frame observes occurring in the moving frame, may be described as rotations in the four-dimensional space-time continuum. From the Minkowski point of view, an interval of time may be regarded as a length in the four-dimensional space-time continuum lying in the t-direction.

The time interval is multiplied by c so that it has the same dimension as a length. Let us reconsider the example of the lightning striking the two ends of a train. From the point of view of the observer at rest, the lightning strikes the front of the train $v/c^2 L_0$ seconds before it hits the end of the train.

The spatial length of the train, on the other hand, is contracted and hence, appears to have the length $L_0 = \sqrt{1 - v^2/c^2}$. The separation of the two events, which to the moving observer were purely spatial, become both spatial and temporal to the stationary observer, demonstrating the equivalence of space and time.

The observation of the stationary observer may be obtained from the moving observer's observations by rotating in the space-time continuum about an axis perpendicular to the x- and t- axes so that a length, which is purely spatial in the primed co-ordinates, becomes both spatial and temporal in the unprimed co-ordinates.

Another example of this rotation in the space-time continuum was encountered when we considered the time dilation of a moving clock. In the frame moving with the clock, the two events of the clock reading 8:00 a.m. and 9:00 a.m., are separated by a purely temporal length.

In the stationary frame, however, the two events are separated by both a spatial and temporal duration, since the clock moves with respect to the stationary observer. Einstein utilized the Minkowski four-dimensional space-time continuum to formulate his theory of gravity. He acknowledged his debt to Minkowksi referring to his contribution as follows:

"Without it, the general theory of relativity would perhaps get no farther than its long clothes".

Although the space-time continuum is four-dimensional in the special theory of relativity, its geometry is still basically Euclidean. Space is not warped or curved. The shortest distance between two points is a straight line, as is the trajectory of a light beam.

In the general theory of relativity, however, the geometry of the space-time continuum is no longer Euclidean; space is curved. The curvature of space was demonstrated, theoretically, using the equivalence principle and, subsequently, verified experimentally, with the detection of the curvature of a beam of light passing close to the sun during a solar eclipse.

The shortest distance between two points in the vicinity of the sun is not a straight line. The most basic axiom of Euclidean geometry, which Kant had elevated to an *a priori* truth, is ironically and empirically untrue. One must describe the four-dimensional space-time continuum in terms of non-Euclidean geometry which, fortunately for Einstein, had been worked out by Lobachevsky, Gauss and Riemann in the nineteenth century.

In non-Euclidean geometry, the ratio of the circumference of a circle to its diameter is no longer equal to pi as it is in Euclidean geometry. We can show that the ratio of the circumference of the diameter of a circle in a rotating frame of reference is less than pi and hence, invoking the equivalence principle demonstrates the need to describe the space-time continuum embedded in a gravitational field in terms of non-Euclidean geometry.

Let us consider a disc which is a perfect circle at rest with the circumference equal to ? Times its diameter. Let us consider the same disc rotating such that its outer edge has the velocity v with respect to an observer sitting at rest at the centre of the disc.

To this observer, a length on the edge of the disc will appear contracted by the Lorentz-Fitzgerald factor $\sqrt{1v^2/c^2}$ because of the motion of the outer edge. This observer will measure the circumference to be $\pi D = \sqrt{1 - v^2/c^2}$.

The diameter of the circumference will not appear contracted, however, because it lies perpendicular to the motion of the disc. The ratio of the circumference to the diameter of a circle in this rotating frame of reference is not but $\pi\sqrt{1 - v^2/c^2}$. The geometry in this frame of reference is non-Euclidean, and since a frame in non uniform motion is equivalent to a frame in which there is a gravitation field, we see that the geometry of the space-time continuum is also non-Euclidean.

Space is curved in a gravitational field. The slowing down of a clock is also easily demonstrated in our rotating frame. A clock at the edge of the disc will appear to slow down to our observer at the centre by virtue of its velocity.

The curvature of space is difficult to comprehend in three dimensions, let alone four dimensions, as required in relativity theory. Let us consider a two-dimensional example to give ourselves a felling for non-Euclidean geometry.

We shall compare a flat, two-dimensional plane, described by

Euclidean geometry, with the surface of a sphere. On the flat plane, the shortest distance between two points is a straight line and the ratio of the circumference to the diameter of a circle is v. The surface of a sphere is a curved, two-dimensional space, described by non-Euclidean geometry.

The shortest distance between two points is not a straight line but a segment of a circle which includes the two points under consideration and whose centre is the centre of the sphere. This circle is called a great circle and the path between the two points is called the geodesic.

In formulating the gravitational interaction between masses, Einstein does not use the concept of one mass exerting a force upon another. Instead, he calculates the curvature of the space-time continuum due to the presence of matter. He then assumes that a mass will travel along the shortest possible path in this non-Euclidean, four-dimensional space. The path, geodesic or world line, along which the particle travels, is determined by the curvature of the space and the initial position and velocity of the particle.

From Einstein's point of view, the earth remains on its world line, not because of the force exerted on it by the sun, but rather because this is the shortest possible path it can find in the curved space-time continuum surrounding the mass of sun. Einstein's model of the gravitational interaction is able to account for something which Newton's model could never explain. Mercury, the closest planet to the sun, orbits the sun every 88 days in an eliptical orbit. The entire orbit of Mercury rotates at an extremely slow rate about the sun.

The distance of closest approach, perihelcon, of Mercury, advances 43 seconds or arc per century. Since there are 360 (in a circle, 60 minutes in a degree and 60 seconds a minute, it would take a little more than 3,000,000 years for the perihelcon of Mercury to make one complete rotation about the sun.

This extremely small effect, nevertheless, cannot be explained in terms of Newton's theory of gravity. Astronomers tried to explain this effect in terms of the interaction of the other planets on Mercury. In fact, the existence of a planet lying between Mercury and the sun, called Vulcan, was postulated to explain the advance of the perihalcon of Mercury.

This planet was never discovered and the anomolous behaviour of Mercury remained a mystery until Einstein's general theory of relativity which was able to explain this effect in terms of the curvature of space. Decke has since shown that the advance for the perihalcon of Mercury could also be explained if the shape of the sun were not spherical.

Since the shape of the sun is not known, this is also a possible explanation. The experimental evidence to support the general theory of relativity is not as conclusive as the evidence supporting the special theory of relativity. Experimental confirmation of Einstein's special theory occurs every day in a laboratory of the elementary particle physicists who explore the properties of particles using high energy accelerators. Experimental confirmation of general relativity has only been obtained in three instances:

1. The bending of light by the sun
2. The shift of spectra from the white-dwarf star and in the laboratory due to a gravitational field.
3. An explanation of the advance of the perihalcon of Mercury.

Research in general relativity continues. Two areas of research that are being actively pursued in general relativity are the search for a unified field theory and the detection of gravity waves. After Einstein published his General Theory of Relativity, he pursued the problem of finding an underlying structure of the gravitational field and the electromagnetic field which would unite them.

In the special theory of relativity, he was able to find the underlying structure which united the electric and magnetic fields. In the general theory of relativity, he showed that gravity may be regarded as a property of the space-time continuum. He tried to show that the electromagnetic field could also be explained as a property of the space-time continuum so he could unite the gravitational and the electromagnetic fields. He never succeeded during his lifetime to solve this problem.

The idea of gravity waves was proposed in analogy to electromagnetic waves. If the electromagnetic field has waves in the form of light, it would seem that the gravitational waves should also exist. Electromagnetic waves are generated through the acceleration of charge, so, perhaps, the acceleration of matter would produce gravity waves. Theoreticians estimated that gravity waves will be very weak. The experimental detection of gravity waves has not been established. Experimental work in this area continues.

KUHN'S STRUCTURE OF SCIENTIFIC REVOLUTIONS AND THE IMPACT OF THE THEORY OF RELATIVITY

Einstein's special and general theories of relativity caused a revolution in scientific thought which affected a number of other fields as well. We will study the nature of this revolution and its influence on the various aspects of human thought. Thomas Kuhn, in his book, *The Structure of Scientific Revolutions,* proposes a model to explain the nature of how a scientific revolution takes place. We shall review Kuhn's theory and examine the Einsteinian revolution in terms of it.

Kuhn does not regard the history of science as the accumulation of "the facts, theories and methods collected in current texts". He believes this view of history arises from the tradition of teaching physics from textbooks in which the true historical processes are suppressed and science is presented as an accumulation of knowledge. Kuhn sees the history of science as a competition between different world views in which revolutions periodically occur whenever a world view fails to accommodate new observations or new ways of looking at older observations.

He claims that: "The early developmental stages of most sciences have been characterized by continual competition between a number of distinct views of nature, each partially derived from, and all roughly compatible with, the dictates of scientific observation and method."

One body of beliefs wins out over the others because it provides a more satisfying picture. This usually occurs as a result of some success that the new theory achieves. This success becomes the model or paradigm for future scientific work which Kuhn labels as normal science. The paradigm becomes a conceptual box into which all of nature is forced to fit. The paradigm reduces the number of things the scientists must consider. He can take the foundation and assumption of his work for granted and can, therefore, exert his full effort in extending, articulating and generalizing his paradigm.

Within the context of normal science, novelty is suppressed. No changes in scientific thinking takes place until new facts arise which can not be accommodated by the old paradigm. Out of the frustration of the failure to fit the new information into the old conceptual box, a new picture emerges. The new paradigm is usually proposed by someone new to the field who has not become set in his ways through frequent use of the old paradigm. With the proposal of a new paradigm a revolution in thinking takes place in which the old view and new view are in conflict and competition.

Eventually, one of the theories triumphs and a return to normal science ensues in which the whole revolutionary process may repeat itself. Kuhn does not regard only the major upheavels in though such as those brought about by Copernicus, Newton or Einstein as the only scientific revolutions. He also regards the discovery of X-rays or Maxwell's formulation of electromagnetic theory as a scientific revolution as well. Each of them brought about a new way of thinking.

Perhaps the most controversial aspect of Kuhn's view of science history is his rather startling claim that the resolution of the conflict between the two competing theories during the revolutionary period is not really rational. The two sides of the controversy make different assumptions, speak a different language and hence, really don't communicate with each other. Max Planck, the man whom the quantum revolution began, perhaps best expressed this idea:

> "A new scientific truth does not triumph by convincing its opponents and making them see the light, but rather because its opponents eventually die, and a new generation grows up that is familiar with it."

An illustration of what Kuhn and Planck are saying is found in the response of William Magie to Einstein's theory of relativity. Magie felt that, although the formulae of Einstein described phenomena accurately from a mathematical point of view, they did not really explain the phenomena. Only an explanation which he could comprehend with his common sense, i.e., fit into his notion of normal science, was valid as far as he was concerned. His attitude is reflected in the following passage from his presidential address to the American Physical Society in 1911:

> "The elements of which the model is constructed must be of types which are immediately percieved by everybody as the ultimate data of consciousness. It is only out of such elements that an explanation, in distinction from a mere barren set of formulae, can be constructed. A description of phenomena, in terms of four dimensions in space, would be unsatisfactory to me as an explanation, because by no stretch of my imagination, can I make myself believe in the reality of a fourth dimension. The description of phenomena in terms of a time, which is a function of the velocity of the body on which I reside, will be, I fear, equally unsatisfactory to me, because, try I ever so hard, I can not make myself realise that such a time is conceivable."

In contrast to the response of Magie, the supporter of the old Newtonian paradigm, we have the euphoric response of Eddington. Eddington was a member of the 1919 solar eclipse expedition to Africa which detected the bending of starlight by the sun. He finds Einstein's new ideas exhilarating and liberating as this is, except from his early writings, reveals:

> "To free our thought from the fetters of space and time is an aspiration of the poet and the mystic, viewed somewhat coldly by the scientist who has too good a reason to fear the confusion of loose ideas likely to ensue. If others have had a suspicion of the end to be desired, it has been left to Einstein to show the way to rid ourselves of those 'terrestrial adhesions to thought.' And in removing our fetters he leaves us, not (as might have been feared) vague generalities for the ecstatic contemplation of the mystic, but a precise scheme of world-structure to engage the mathematical physicist."

Kuhn maintains that a scientific revolution occurs when the existing forms can no longer accommodate new facts and offer no possibility of resolution. The only recourse is to a new system with a different set of values. It is obvious that Magie and Eddington have a different set of values and hence, can never convince each other who is correct. It can be argued that Newtonian physics is not really incompatible with Einsteinian physics since, as we saw earlier, that the relativistic effects disappear as the velocities involved become very small compared with the velocity of light. While it is true that the formulae become identical in the limit of small velocities, the interpretation of space and time still remains radically different for the Newtonian and the Einsteinian.

Because of the fact that the new theory always incorporates the valid results of the old theory, science continues to make progress by definition. Kuhn claims that the progress made by science arises from the fact that the range of phenomena explained is continually increasing, not because the Einsteinian point of view is superior to the Newtonian point of view. There are no absolute truths for Kuhn. In this sense, his ideas have been influenced by Einstein's theory of relativity. Kuhn points out that scientific revolutions are often fomented by those who are new to the field and/or those who do not belong to the scientific establishment. No one could provide a better example of this viewpoint than Einstein as the following brief sketch of his life reveals.

Einstein was born in Bavaria in 1879 to a middle class Jewish family. He was a slow learner as a child. His parents feared that he was retarded because he was so late in learning how to speak. He was never a good student since he was given to daydreaming. At the age of fifteen, he dropped out of school and travelled in Italy. He finally settled down, applied for admission to university, and failed the entrance exam. He went back to secondary school for a year and passed the entrance exam to the Swiss Federal Polytechnical School in Zurich. He was not a model student in university, either. He passed his exams by cramming his friends' notes. He preferred reading on his own to attending lectures. He found the whole experience of higher education appalling as the following remarks penned some years later reflect:

> "... after I had passed the final examination, I found the consideration of any scientific problem distasteful to me for an entire year. It is little short of a miracle that modern methods of instruction have not already completely strangled the holy curiosity of inquiry, because what this delicate little plant needs most, apart from initial stimulation, is freedom; without that it is surely destroyed... I believe that one could even deprive a healthy beast of prey of its voraciousness, if one could force it with a whip to eat continuously whether it were hungry or not..."

He, nevertheless, completed his Ph.D. in physics. Afterwards, he earned money as a tutor until an influential friend found him an undemanding job in the Swiss Patent Office on Berne. He held this position for seven years (1902-1909), during which he published his most famous papers. In 1905, the year his papers on the Special Theory of Relativity appeared, he also published two other major works, which, in themselves, also represented revolutions in physics thinking. It was for his 1905 paper on the photoelectric effect, which established the existence of the photon, and not relativity that he received the Nobel Prize in 1921. In the same year, he published his paper describing Brownian motion which established the first direct detection of atoms.

It was in 1909, after he had completed his most revolutionary work with the exception of the general theory of relativity, that Einstein began to receive academic acclaim. In the years following 1909, he accepted professional chairs from Prague, Zurich and Berlin. In 1932, as Hitler came to power in Germany, Einstein fled to the United States where he spent the remainder of his life. He was also a great humanist devoting much of his efforts to world peace. Einstein's theory of relativity caused as large a response in the non-scientific world as it did in the scientific world. The overwhelming response of the layman was bewilderment because the notions of relativity violated their notion of common sense very much as it had violated the common sense of the physicist, Magie. Relativity for them was strictly a mathematical model which they could not relate to intuitively.

One aspect of Einstein's ideas, however, became very popular and that was the notion of relativity. If everything is relative in the physical world, it was argued then that the same must be true of things in the world of art, morals or ideas. Some people were appalled by this notion and were opposed to Einstein's ideas because they considered them downright immoral. Other thinkers found the aspect of Einstein's physical theories extremely liberating. For example, Jose Ortega y Gasset wrote in his book, *The Modern Theme*;

"The theory of Einstein is a marvelous proof of the harmonious multiplicity of all possible points of view if the idea is extended to morals and aesthetics, we shall come to experience history and life in a new way. It is the same with nations. Instead of regarding non-European cultures as barbarous, we shall now begin to respect them, as methods of confronting the cosmos which are equivalent to our own. There is a Chinese perspective which is fully as justified as the Western."

Another thinker, Paul Squires, saw in Einsteinian relativity support for the notion of gestalt psychology. He likens traditional psychology to Newtonian Physics and Gestalt Psychology to Einsteinian physics. He writes:

> "For the traditional psychology, particular experiences and bets of behaviour possess a more or less absolute character. For the Gestalt psychologist, any aspect of mentality has meaning only in its relation to the larger context, the whole, in which it exists."

The art critic, Thomas Craven, in an article written in 1921, claimed that Einstein's ideas support the position of the modern artists. He wrote:

> "While the celebrated physicist has been evolving his shocking theories of the course of natural phenomena, the world of art has suffered an equivalent heterodoxy with respect to its expressive media. This revolt has sprung from concoction that the old art is not necessarily infallible, and that equally significant achievements may be reached by new processes and by fresh sources of inspiration."

Craven regards the rigidity of "the old art" as corresponding to the "immovable reference-body" of Newtonian physics. He claims just as Einstein's theory changed Newton's concepts of absolute space and time "similarly has the modern painter broken the classical tradition." Rather than clinging to the rigid laws of photographic vision, he claims that the artist injects his own "personal feeling". Instead of arriving at an absolute truth "which serves all purposes of illustration" but "reveals nothing psychologically", the artist achieves at to greater truth, a psychological truth, relative to himself.

Chapter 9

Radioactivity

In radioactive processes, particles or electromagnetic radiation are emitted from the nucleus. The most common forms of radiation emitted have been traditionally classified as alpha (α), beta (β), and gamma (γ) radiation. Nuclear radiation occurs in other forms, including the emission of protons or neutrons or spontaneous fission of a massive nucleus.

Of the nuclei found on Earth, the vast majority are stable. This is so because almost all short-lived radioactive nuclei have decayed during the history of the Earth. There are approximately 270 stable isotopes and 50 naturally occurring *radioisotopes* (radioactive isotopes). Thousands of other radioisotopes have been made in the laboratory.

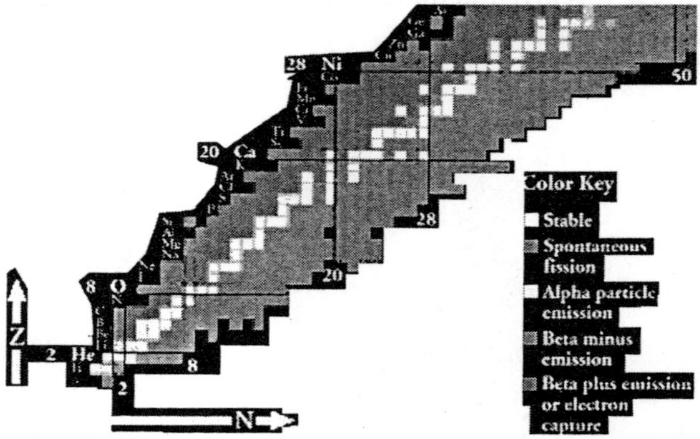

Radioactive decay will change one nucleus to another if the product nucleus has a greater nuclear binding energy than the initial decaying nucleus. The difference in binding energy (comparing the before and after states) determines which decays are energetically possible and which are not. The excess binding energy appears as kinetic energy or rest mass energy of the decay products.

The Chart of the Nuclides, part of which is shown above is a plot of nuclei as a function of proton number, Z, and neutron number, N. All stable nuclei and known radioactive nuclei, both naturally occurring and manmade, are shown on this chart, along with their decay properties.

Nuclei with an excess of protons or neutrons in comparison with the stable nuclei will decay toward the stable nuclei by changing protons into neutrons or neutrons into protons, or else by shedding neutrons or protons either singly or in combination. Nuclei are also unstable if they are excited, that is, not in their lowest energy states. In this case the nucleus can decay by getting rid of its excess energy without changing Z or N by emitting a gamma ray.

Nuclear decay processes must satisfy several conservation laws, meaning that the value of the conserved quantity after the decay, taking into account all the decay products, must equal the same quantity evaluated for the nucleus before the decay. Conserved quantities include total energy (including mass), electric charge, linear and angular momentum, number of nucleons, and *lepton number* (sum of the number of electrons, neutrinos, positrons and antineutrinos–with antiparticles counting as -1).

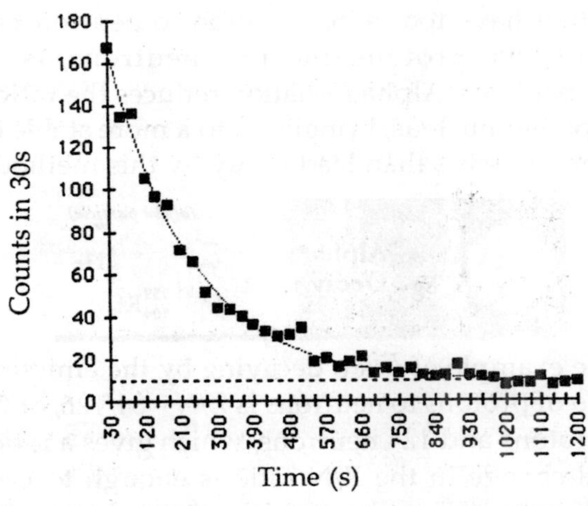

137Ba decay data, counting numbers of decays observed in 30-second intervals. The best-fit exponential curve is shown. The points do not fall exactly because of statistical counting fluctuations.

The probability that a particular nucleus will undergo radioactive decay during a fixed length of time does not depend on the age of the nucleus or how it was created. Although the exact lifetime of one particular nucleus cannot be predicted, the mean (or average) lifetime of a sample containing many nuclei of the same isotope can be predicted and measured. A convenient way of determining the lifetime of an isotope is to measure how long it takes for one-half of the nuclei in a sample to decay–this quantity is called the *half-life*, $t_{1/2}$. Of the original nuclei that did not decay, half will decay if we wait another half-life, leaving one-quarter of the original sample after a total time of two half-lives. After three half-lives, one-eighth of the original sample will remain and so on. Measured half-lives vary from tiny fractions of seconds to billions of years, depending on the isotope.

The number of nuclei in a sample that will decay in a given interval of time is proportional to the number of nuclei in the sample. This condition leads to radioactive decay showing itself as an exponential process, as shown above. The number, *N*, of the original nuclei remaining after a time *t* from an original sample of N_0 nuclei is

$$N = N_0 e^{-(t/T)}$$

where *T* is the mean lifetime of the parent nuclei. From this relation, it can be shown that $t_{1/2} = 0.693T$.

ALPHA DECAY

An alpha-particle decay: In alpha decay, shown in the figure, the nucleus emits a 4He nucleus, an *alpha particle.* Alpha decay occurs most often in massive nuclei that have too large a proton to neutron ratio. An alpha particle, with its two protons and two neutrons, is a very stable configuration of particles. Alpha radiation reduces the ratio of protons to neutrons in the parent nucleus, bringing it to a more stable configuration. Many nuclei more massive than lead decay by this method.

Consider the example of ^{210}Po decaying by the emission of an alpha particle. The ratio of protons to neutrons is $Z/N = 84/126$, or 0.667. A 206Pb nucleus has 82 protons and 124 neutrons, which gives a ratio of 82/124, or 0.661. This small change in the Z/N ratio is enough to put the nucleus into a more stable state, and as shown in the figure, brings the "daughter" nucleus (decay product) into the region of stable nuclei in the Chart of the Nuclides.

In alpha decay, the atomic number changes, so the original (or parent) atoms and the decay-product (or daughter) atoms are different elements and therefore have different chemical properties.

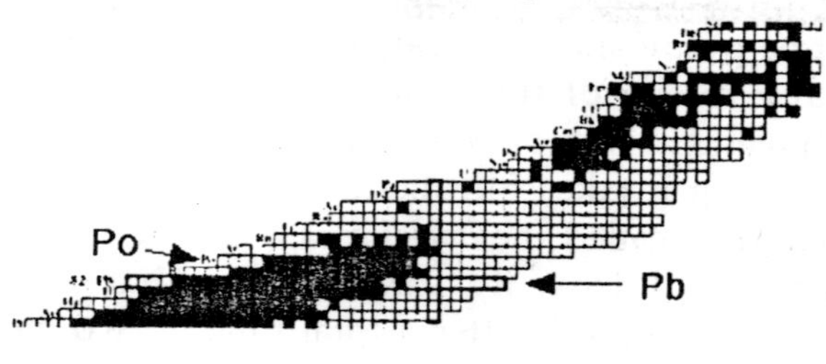

Upper end of the Chart of the Nuclides: In the alpha decay of a nucleus, the change in binding energy appears as the kinetic energy of the alpha particle and the daughter nucleus. Because this energy must be shared between these two particles, and because the alpha particle and daughter nucleus must have equal and opposite momenta, the emitted alpha particle and recoiling nucleus will each have a well-defined energy after the decay. Because of its smaller mass, most of the kinetic energy goes to the alpha particle.

BETA DECAY

Beta decays: Beta particles are electrons or *positrons* (electrons with positive electric charge, or *antielectrons*). Beta decay occurs when, in a nucleus with too many protons or too many neutrons, one of the protons or neutrons is transformed into the other. In beta minus decay, a neutron decays into a proton, an electron, and an antineutrino. In beta plus decay, a proton decays into a neutron, a positron, and a neutrino. Both reactions occur because in different regions of the Chart of the Nuclides, one or the other will move the product closer to the region of stability.

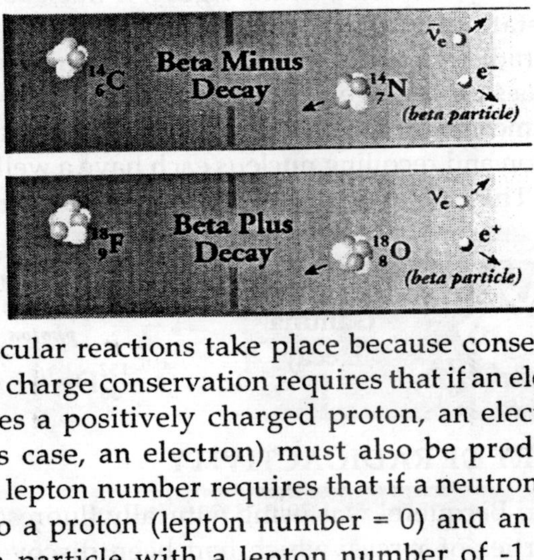

These particular reactions take place because conservation laws are obeyed. Electric charge conservation requires that if an electrically neutral neutron becomes a positively charged proton, an electrically negative particle (in this case, an electron) must also be produced. Similarly, conservation of lepton number requires that if a neutron (lepton number = 0) decays into a proton (lepton number = 0) and an electron (lepton number = 1), a particle with a lepton number of -1 (in this case an antineutrino) must also be produced. The leptons emitted in beta decay did not exist in the nucleus before the decay–they are created at the instant of the decay.

To the best of our knowledge, an isolated proton, a hydrogen nucleus with or without an electron, does not decay. However within a nucleus, the beta decay process can change a proton to a neutron. An isolated neutron is unstable and will decay with a half-life of 10.5 minutes. A neutron in a nucleus will decay if a more stable nucleus results; the half-life of the decay depends on the isotope. If it leads to a more stable nucleus, a proton in a nucleus may capture an electron from the atom (electron capture), and change into a neutron and a neutrino.

Proton decay, neutron decay, and electron capture are three ways in which protons can be changed into neutrons or vice-versa; in each decay there is a change in the atomic number, so that the parent and daughter atoms are different elements. In all three processes, the number *A* of nucleons remains the same, while both proton number, *Z*, and neutron number, *N*, increase or decrease by 1.

In beta decay the change in binding energy appears as the mass energy and kinetic energy of the beta particle, the energy of the neutrino, and the kinetic energy of the recoiling daughter nucleus. The energy of an emitted beta particle from a particular decay can take on a range of values because the energy can be shared in many ways among the three particles while still obeying energy and momentum conservation.

GAMMA DECAY

In gamma decay, depicted in the figure, a nucleus changes from a higher energy state to a lower energy state through the emission of electromagnetic radiation (*photons*). The number of protons (and neutrons) in the nucleus does not change in this process, so the *parent* and *daughter* atoms are the same chemical element. In the gamma decay of a nucleus, the emitted photon and recoiling nucleus each have a well-defined energy after the decay. The characteristic energy is divided between only two particles.

THE DISCOVERY OF RADIOACTIVITY

In 1896 Henri Becquerel was using naturally fluorescent minerals to study the properties of *x-rays*, which had been discovered in 1895 by Wilhelm Roentgen. He exposed potassium uranyl sulfate to sunlight and then placed it on photographic plates wrapped in black paper, believing that the uranium absorbed the sun's energy and then emitted it as x-rays. This hypothesis was disproved on the 26th-27th of February, when his experiment "failed" because it was overcast in Paris. For some reason, Becquerel decided to develop his photographic plates anyway. To his surprise, the images were strong and clear, proving that the uranium emitted radiation without an external source of energy such as the sun. Becquerel had discovered radioactivity.

Becquerel used an apparatus similar to that displayed below to show that the radiation he discovered could not be x-rays. X-rays are neutral and cannot be bent in a magnetic field. The new radiation was bent by the magnetic field so that the radiation must be charged and different than x-rays. When different radioactive substances were put in the magnetic field, they deflected in different directions or not at all, showing that there were three classes of radioactivity: negative, positive, and electrically neutral.

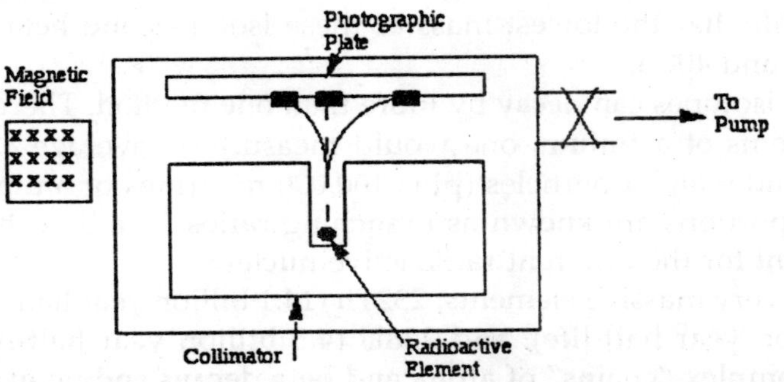

The term radioactivity was actually coined by Marie Curie, who together with her husband Pierre, began investigating the phenomenon recently discovered by Becquerel. The Curies extracted uranium from ore and to their surprise, found that the leftover ore showed more activity than the pure uranium.

They concluded that the ore contained other radioactive elements. This led to the discoveries of the elements polonium and radium. It took four more years of processing tons of ore to isolate enough of each element to determine their chemical properties.

Ernest Rutherford, who did many experiments studying the properties of radioactive decay, named these alpha, beta, and gamma particles, and classified them by their ability to penetrate matter. Rutherford used an apparatus similar to that depicted in the figure.

When the air from the chamber was removed, the alpha source made a spot on the photographic plate. When air was added, the spot disappeared. Thus, only a few centimeters of air were enough to stop the alpha radiation.

Because alpha particles carry more electric charge, are more massive, and move slowly compared to beta and gamma particles, they interact much more easily with matter. Beta particles are much less massive and move faster, but are still electrically charged. A sheet of aluminum one millimeter thick or several meters of air will stop these electrons and positrons. Because gamma rays carry no electric charge, they can penetrate large distances through materials before interacting–several centimeters of lead or a meter of concrete is needed to stop most gamma rays.

RADIOACTIVITY IN NATURE

Radioactivity is a natural part of our environment. Present-day Earth contains all the stable chemical elements from the lowest mass (H) to the highest (Pb and Bi). Every element with higher Z than Bi is radioactive. The earth also contains several primordial long-lived radioisotopes that have survived to the present in significant amounts. 40K, with its 1.3 billion year half-life, has the lowest mass of these isotopes and beta decays to both 40Ar and 40Ca.

Many isotopes can decay by more than one method. Therefore from 100,000 atoms of actinium, one would measure on average 83,000 beta particles and 6 alpha particles (plus 100,000 neutrinos or antineutrinos). These proportions are known as branching ratios. The branching ratios are different for the different radioactive nuclei.

Three very massive elements, 232Th (14.1 billion year half-life), 235U (700 million year half-life), and 238U (4.5 billion year half-life) decay through complex "chains" of alpha and beta decays ending at the stable 208Pb, 207Pb, and 206Pb respectively. The *decay chain* for 238U is shown in the figure.

The ratio of uranium to lead present on Earth today gives us an estimate of its age (4.5 billion years). Given Earth's age, any much shorter lived radioactive nuclei present at its birth have already decayed into stable elements. One of the intermediate products of the ^{238}U decay chain, ^{222}Rn (radon) with a half-life of 3.8 days, is responsible for higher levels of background radiation in many parts of the world. This is primarily because it is a gas and can easily seep out of the earth into unfinished basements and then into the house.

Some radioactive isotopes, for example 14C and 7Be, are produced continuously through reactions of cosmic rays (high energy charged particles from outside Earth) with molecules in the upper atmosphere. 14C is useful for radioactive dating. Also, the study of radioactivity is very important to understand the structure of the earth because radioactive decay heats the earth's interior to very high temperatures.

The uranium decay series. The vertical axis is atomic mass.

UNITS OF RADIOACTIVITY

The number of decays per second, or *activity*, from a sample of radioactive nuclei is measured in *becquerel* (Bq), after Henri Becquerel. One decay per second equals one becquerel.

An older unit is the *curie*, named after Pierre and Marie Curie. One curie is approximately the activity of 1 gram of radium and equals (exactly) 3.7×10^{10} becquerel. The activity depends only on the number of decays per second, not on the type of decay, the energy of the decay products, or the biological effects of the radiation.

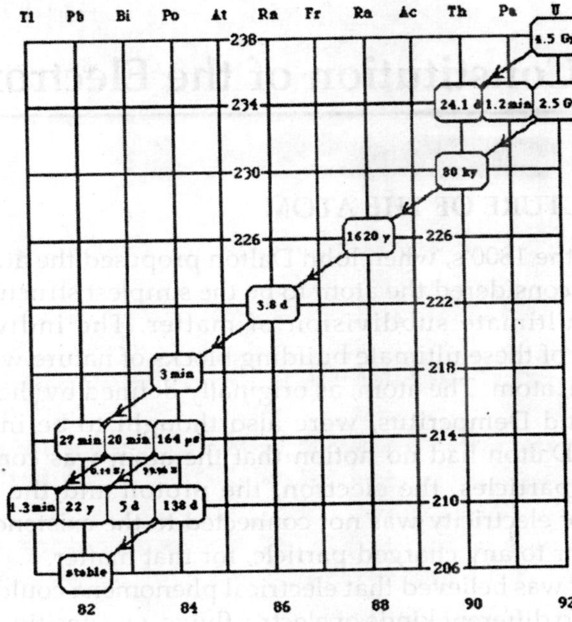

Chapter 10

Constitution of the Electron

THE STRUCTURE OF THE ATOM

Early in the 1800's, when John Dalton proposed the atomic structure of matter; he considered the atom to be the simplest structure possible in nature, the ultimate subdivision of matter. The indivisibility and immutability of these ultimate building blocks of nature were implied in the very term, atom. The atom, as originally defined by the early Greeks, Leucippus and Democritus, were also thought to be indivisible and unchanging. Dalton had no notion that the atom was composed of the still smaller particles, the electron, the proton and the neutron. The phenomena of electricity was not connected to the existence of electrons and protons or to any charged particle, for that matter.

Rather, it was believed that electrical phenomena could be explained in terms of two different kinds of electric fluids, one positive and the other negative. The term current itself, used to describe a flow of electrically charged particles, retains the original connotation of the fluid nature of electricity. The electric fluid used to describe charge was considered to be massless and was not associated with matter or atoms. It was an independent constituent of nature.

The laws of electricity and magnetism were originally formulated in terms of the electric fluids rather than in terms of charged particles. All of the notions of electromagnetic phenomena, including Maxwell's equations, were conceived in terms of the density and the current of the positive and negative electric fluids. It was not until after the electromagnetic interaction was understood that it was realized that the density of the two electric fluids was the density of electrons and protons and that currents were generated by the movement of these particles.

Just as the notion of caloric fluid disappeared once it was realized that heat was due to the motion of the atom, so, too, the notion of electric fluids disappeared once it was realized that electric phenomena were due to the underlying electrical structure of atoms.

We now turn to a description of the experiments which led to the discovery of the underlying electrical structure of the atom. Students are often baffled as to how the physicists know that an atom consists of a positively charged nucleus about which negatively charged electrons orbit.

They become skeptical about the existence of the electron when they discover that the electron has not and can not be visually observed. In order to counteract this skepticism, we shall carefully describe a series of experiments stretching over a period of almost 100 years in which the existence, charge and mass of the electron were established and in which the underlying structure of the atom was discovered.

The first hint of any connection between the atom and electrical phenomena came from the observation of the process of electrolysis. We have illustrated a typical electrolysis experiment. The anode or positive electrode and the cathode or the negative electrode are two metal rods connected to the positive and negative terminals of a battery and are immersed in water.

Note the presence of an a meter in the circuit placed between the negative terminal of the battery and the cathode. This device permits one to measure the amount of electricity flowing in the circuit. As the current flows in this circuit, hydrogen and oxygen gas will collect at the cathode and anode respectively. The volume of hydrogen gas that collects is twice that of the oxygen gas and is proportional to the total amount of electricity that flows in the circuit.

The flow of electricity is possible because the water molecules, H_20, separate chemically into two positively charged hydrogen ions, H^+ and an oxygen ion, O^{--}. An ion is an atom which carries a charge. Because there is an excess of negative charge at the cathode, the H^+ ions collect there, become neutralized and form hydrogen gas. The oxygen ions drawn by the positive charge collect at the anode, are neutralized and form oxygen gas. The ratio of the volume of hydrogen gas to oxygen gas is two to one because there are two hydrogen atoms to one oxygen atom in the water molecule.

Other electrolysis experiments showed that most atoms are capable of carrying a charge. One such experiment led to the process of silver plating. This time, however, we choose, for the anode, a pure silver rod, and for the cathode, a copper rod.

Instead of water as our conducting medium, we choose a solution of silver nitrate, A_gNO_3. When silver nitrate is dissolved in water, silver ions, $A_g{}^+$, are attracted to the copper cathode to which they attach themselves forming a silver plating on the cathode. The nitrate ions, $NO_3{}^-$, are attracted to the silver anode where they combine chemically with the silver to produce more silver nitrate.

The level of silver nitrate remains the same as a result of this process, silver, however, is transferred from the silver anode to the copper cathode where it forms a uniform plate. One can reverse this process and remove the silver plate from the copper rod by changing the direction of the current. This is easily accomplished by switching the terminals of the battery.

The electrolysis experiments first performed by Humphrey Davy and Michael Faraday on the early part of the nineteenth century were motivated by a desire to study the passage of an electric current through a liquid. Faraday, during this period, also began to study the problem of the passage of an electric current through a gas. He discovered that gases are excellent insulators and that the conductivity of the gas increases as its density decreases.

Faraday's research was limited by the primitive state of vacuum techniques, at the time. It was not until 1854 when Heinrich Geissler invented a vacuum pump which enabled him to evacuate a glass tube that the passage of an electric current through a gas could be properly studied.

A device similar to the Geissler tube. The cathode and anode are attached respectively to the negative and positive terminal of a high voltage battery creating a strong electric field inside the discharge tube. The passage of an electric current occurs as a result of the ionization of the gas molecules (or atoms), the process whereby an electron is removed from the gas molecule (or atom) leaving a positively charged ion.

At the time these experiments were first performed, it was thought that the ionization was due to the strength of the electric field in the tube which was thought to tear apart the molecules of the gas. We have since discovered that the ionization is due to the effects of cosmic radiation, the high energy charged particles, x-rays and gamma rays that are continually streaming towards earth from outer space.

Once the molecule is ionized, the positive ion begins to accelerate the anode, each gaining kinetic energy from the electric field. If the gas is dense, they will soon collide with a neutral molecule, lose their kinetic energy, and begin accelerating once again toward their respective electrode.

In this way, a small current is able to pass through a dense gas. As the tube is evacuated so that the density of the gas becomes significantly lower, the charge particles do not collide as often with the neutral molecules of the gas. They, therefore, have time to build up more kinetic energy so that when they finally collide with a neutral molecule, they can ionize it creating another electron-ion pair. The new ion pair will then go on to create many more additional ion pairs.

This process snowballs and an electric discharge between the cathode and the anode takes place, an appreciable current begins to flow, and the tube begins to glow visibly. Not all of the electrons and ions in the gas actually reach the anode or the cathode.

Some of them recombine on the way with new partners. Because of the kinetic energy the ion pair has acquired from the electric field, the neutral molecule formed through the recombination is often in an excited state. It rids itself of this excess energy by radiating visible light which partly accounts for the glow from the centre of the tube.

There is also a glowing of the glass at either end of the tube which is due to the stream of charged particles. When this glow was first observed, it was not understood what caused it. It was thought to be some kind of radiation.

The glow at the anode caused by the electrons, for example, was attributed to cathode rays which presumably emenated from the cathode. By cutting a hole in the electrodes, William Crookes was able to study the nature of the radiation causing the glass of the tube to glow. William Crookes showed that the location of the glow could be displaced by a magnet. He cited this as evidence that the so-called radiation was actually due to a stream of charged particles.

This work was corroborated and further developed by J.J. Thomson in 1897, who showed that the stream of particles could also be deflected by an electric field.

By comparing the deflections due to the electric and magnetic fields, Thomson was able to determine the ratio of charge to mass for the charged particles. His results of other determinations of the charge to mass ratio of heavy ions obtained with electrolysis studies. His results with the negatively charged particles were extremely surprising. He found that the ratio of charge to mass of these negatively charged particles was thousands of times larger than the same ratio of the negative ions observed in electrolysis.

If one assumed that the magnitude of these newly discovered particle's charge was the same as that of the hydrogen ion, then, its mass would be only 1/2000 the mass of the hydrogen atom. Thomson associated this particle with the electron, named in 1891 by George Stone, whose existence was postulated as the ultimate particle of electricity. Thomson's result established the electrical composition of the atom.

Soon after Thomson's discovery of the electron, two models of the atom arose. According to the proponents of one point of view, the atom was a miniature solar system. There was a positively charged nucleus at the centre of the atom around which orbited the negatively charged electrons. This model, which eventually developed into our present-day notion of the atom, was in fierce competition with a model proposed by

Thomson himself and which he called "the plum pudding model" of the atom. He assumed that the atom was a sphere of positive charge in which the electrons were embedded like plums in a pudding. In both models, the atom was basically neutral. The atom became a positive or negative ion as electrons were either lost or gained. Both models explained the results of the electrolysis and gas discharge experiments.

The plum pudding atom, however, gained favour over the miniature solar system atom because it provided a more satisfactory description of the electromagnetic radiation emitted by matter. The light emitted by various materials was examined by passing it through a prism before it fell upon a photographic plate.

This enabled the experimentalists to determine the spectral distribution of the being radiated by a particular material. They found that the spectra from the closely-packed solids and liquids are continuous, but, that the spectra from the gases are distinguished by discreet lines. Apparently, the continuous spectra was due to the external motion of the atom as a whole, due to heat. This was verified by the fact that the distribution of frequencies changed as the temperature, the higher the frequency of the radiation.

This is to be expected since the frequency of the light emitted, according to Maxwell's theory of radiation, is related to the frequency of the charged particles emitting the radiation. As the solid or liquid is heated, the atoms move faster and, hence, one expects a shift to higher frequencies. Both the plum pudding and solar system models of the atom could account for the continuous spectrum of radiation.

The spectra from the gases was not as easily explained. Each gas had its own characteristic set of discreet lines implying that this radiation was due to internal motion within the atom. This hypothesis was confirmed by the fact that the heating of the gas did not change these characteristic frequencies. They were unaffected by the external motion of the atom and, hence, were due to the internal motion of the atom.

The plum pudding model explained these discreet lines as due to the oscillations of the electrons in the pudding of positive charge. Proponents of the solar system model tried to explain these frequencies in terms of the frequency of the electron's orbit about the nucleus.

According to Maxwell's theory, however, a charged particle moving in a circular orbit is constantly accelerating and, hence, would be continuously radiating light. As the electron continued to radiate light, it would lose energy, and, hence, fall into an orbit closer to the nucleus where it would lose more energy and so on until it spiralled into the nucleus. This defect of the solar system model of the atom, known as the "spiral death", caused it to fall into disfavour. The solar system atom would arise again and triumph but, for the time being, the plum pudding model provided the best description then of the atom.

Although Thomson's gas discharge experiments had established the existence of the electron, the charge of the electron had never been measured nor had the hypothesis that the electron was the ultimate unit of charge been verified. It wasn't until 1909 that Robert Milikan completed the work begun twelve years earlier by Thomson with his famous oil drop experiment.

Milikan sprayed oil drops into a device, through the narrow nozzle of an atomizer. Many of the oil drops became charged in this process losing or gaining electrons through the act of subdividing or by rubbing against the inner wall of the nozzle or each other.

After spraying the oil drops into the device, he would observe them falling under the influence of gravity. By measuring their terminal velocity, he was able to determine the mass of the oil drop. He would then apply a known electric field between the plates which would accelerate the oil drops upwards. By measuring the rate of the upward acceleration, Milikan was able to determine the absolute charge on an oil drop.

He found that the charge on an oil drop came in discreet units, labeled e. This unit, e, is the absolute charge of the electron and is a fundamental constant of nature. It is an extremely small number equal to 1.6×10^{-19} coulombs. [A coulomb is the amount of charge that flows through a standard 100 watt light bulb in approximately one second.]

The smallest charge observed for an oil drop was e. The charges on the other oil drops were always equal to an integer times e. Thus, one observed oil drops with charge +e, -e, 17e, -10e, +3e, etc., but never with charges like -.5e, 1.1e or 2.37e. Milikan concluded that an electron carries the charge -e and that the charge on his oil drops was due to an excess or lack of a given number of electrons. The oil drop with charge +17e was missing 17 electrons whereas the oil drop with charge -10e had 10 extra electrons.

Once the charge of the electron was known, one was able to determine the mass of the electron using Thomson's determination of the charge to mass ratio. One discovers, in this way, that the electron is an enormously small object with a mass of only 9.1×10^{-28} grams.

Exploiting Thomson's determination of the charge to mass ratio of the hydrogen ion, H+, and assuming that it has the charge +e, we find that the mass of the hydrogen atom is 1.66×10^{-24} grams. Once the mass of the hydrogen atom is determined, the absolute masses of all the other atoms are determined also since the relative mass of the atoms is known from the work of the chemists of the first half of the nineteenth century. The oxygen atom is approximately 16 times the mass of the hydrogen atom and the carbon atom is 12 times the mass of the hydrogen atom. From the absolute weights of the atoms, we can determine the number of hydrogen, carbon or oxygen atoms in one gram of hydrogen, 12 grams of carbon or 16 grams of oxygen, respectively.

This number, which is called Avogadro's number, is equal to 6×10^{23}. Making use of our knowledge of the density of graphite, which is pure carbon, we can determine the volume occupied by a carbon atom and, hence, its radius which turns out to be approximately 10^{-8} cm.

Actually, Avogadro's number was not originally determined by Milikan's experiment of 1909 but by Einstein's analysis of Brownian motion in 1905, Brownian motion, first observed in 1827 by the botanist Robert Motion, is the observed erratic, jerky motion of microscopic pollen grains suspended in water.

The smaller the pollen grain, the more violent is the motion of the grain. Einstein pointed out in 1905 that this jerky motion is due to the fact that the pollen grain is being constantly bombarded by water molecules. From his detailed calculations of the pollen grain's motion, Einstein was able to determine the absolute weight of various atoms and, hence also Avogadro's number. His analysis of Brownian motion actually provided the first direct detection of the atom.

Additional evidence for the composite nature of the atom was gathered towards the end of the nineteenth century as new forms of radiation were discovered.

Rontgen discovered x-rays in 1895 when working with his gas discharge tube, he noticed a glow in some fluorescent material lying near his apparatus. These rays were produced as a result of the bombardment of the atoms of the anode by electrons being accelerated there by the electric field in the gas discharge tube. As a result of the bombardment by the electrons, the atoms in the anode were excited and, as a consequence, emitted x-rays.

It was not until the Van Laue experiments of 1912 that it was realized that x-rays are electromagnetic radiation of extremely small wavelength. The wavelength of x-rays are of the order of 10^{-8} cm., the distance between atoms, which explains their ability to penetrate solid matter. Van Laue passed an x-ray beam through a crystal in which the atoms are arranged in an orderly array.

He found that the x-rays displayed the characteristic diffraction pattern one observes for light passing through a system of slits. The spaces between the atoms in the crystal served as the system of slits. From the diffraction pattern and the knowledge of the spacing between the atoms in the crystal, other investigators were able to determine the wavelength of the x-ray. It was discovered that the x-rays produced by a particular atom had certain discreet characteristic frequencies analogous to the situation with the discreet lines of visible light emitted by gas atoms.

Roentgen's discovery of x-rays stimulated other researchers to look for still different forms of radiation. In 1896, Becquerel began a series of experiments with phoshorescent materials such as the uranium compound, potassium uranyl sulfate, which glows in the dark after it is exposed to sunlight.

Becquerel was hoping to show that the radiation from phosphorescent materials contained x-rays. Quite by accident, he made another discovery which turned out to be far more important. Becquerel left a sample of the potassium uranyl sulfate in his desk drawer with some photographic film and discovered, much to his surprise, that, although no light had fallen upon the film, it was still exposed.

Apparantly, potassium uranyul sulfate emits radiation on its own without being exposed to the sunlight. Becquerel isolated the source of the spontaneous of radiation and found that it was due to uranium. Becquerel's effort led to the work of Pierre and Marie Curie who went on to investigate the nature of radioactivity (the spontaneous emission of radiation from matter, a term coined by Marie Curie) and to isolate a number of radioactive elements such as polonium, thorium and radium.

Subsequent investigation revealed that radioactive atoms emitted three forms of radiation which were labelled alpha rays, beta rays and gamma rays. Beta rays were identified by Becquerel as high energy electrons.

The gamma ray was discovered by P. Villard and identified as electromagnetic radiation with a wave length even shorter than that of x-rays. Alpha rays were discovered by the Curies and Earnest Rutherford and were identified as a stream of particles with the charge, +2e, and a mass four times that of the hydrogen atom. In other words, the alpha particle is nothing more than the nucleus of the helium atom without its two electrons.

The ultimate source of radioactivity within the atom is the nucleus and we shall, therefore, defer our discussion of radioactivity until we come to our treatment of nuclear physics. The importance of radioactivity for the study of atomic physics is twofold. Firstly, it provides, in itself, additional evidence for the composite nature of the atom.

Secondly, and perhaps more importantly, Rutherford realized that, because of its charge, alpha rays could be used as a tool to probe the structure of the atom. He and his colleagues, Marsden and Geiger, devised a scattering experiment in which a collimated beam of alpha particles was directed at a thin gold foil target. In an experiment, the alpha rays originated at the source and pass through a slit in a thick sheet of lead in order to produce a collimated beam. The collimated beam then passes through the gold foil where it is scattered by the individual gold atoms. The scattered alpha rays are then detected with a fluorescent screen to which is attached an eye piece. Whenever an alpha particle strikes the fluorescent screen, a spark of light is produced which is visible through the eyepiece. The fluorescent screen and eyepiece can be moved about the circle so that the number of alpha particles scattered at each angle can be counted.

When Rutherford and his colleagues performed their experiment, they believed that they would establish the experimental validity of J.J. Thomson's plum pudding picture of the atom in which a number of electrons floated in a spherical blob of positive charge with a radius of approximately 10^{-8} cm.

They, therefore, expected that the alpha particle would pass through the atom without scattering very much because the electrons were too light to effect the alpha particles very much and because the positive charge was spread out over the whole atom. The alpha particle was expected to be only nominally scattered from its original path.

Imagine what a surprise it must have been for Rutherford and his co-workers when they discovered that a number of alpha particles suffered considerable deflections and, in fact, some alpha particles had actually been scattered completely into the backwards direction, i.e., scattered back into the direction from which they originally came.

The observed scattering of the alpha particles could only have been caused by the positive charge of the nucleus since the electrons do not have enough mass to scatter the alpha rays very effectively. But, if the positive charge of the nucleus were responsible for the large angle scattering of the alpha particles, then, the positive charge had to be concentrated into a considerably smaller space. In fact, Rutherford showed that the positive charge had to be concentrated in a nucleus with a radius of less than 10^{-12} cm.

The picture of the atom completely changed overnight. The miniature solar system atom came back into favour. The atom now consisted of a nucleus which contains all of the positive charge in a sphere less than 10^{-12} cm in radius. (It is now known that the radius of the nucleus varies from 1 to 7×10^{-13} cm.) The electron circles the nucleus with an orbit whose radius is of the order of 10^{-8} cm. Most of the atom consists of empty space. This was the only picture of the atom consistent with the result of Rutherford's experiment.

This picture of the atom presented a rather grave problem, however. According to the classical theory of electromagnetic theory, any particle accelerating will radiate light energy. Since the electron orbiting the nucleus is continually accelerating, it should also be continually radiating light and hence, losing energy. As it loses energy, it spirals into the nucleus until it is finally absorbed.

Essentially, the miniature solar system atom is unstable from the point of view of classical electromagnetic theory. The solution of this problem which Bohr proposed two years later in 1913, is related to the quantization of the energy of light discovered earlier through the work of Max Planck in 1900 and Albert Einstein in 1905. Before continuing with our story of the atom, we turn our attention to the discovery of the photon and the quantization of energy.

BOHR'S ATOM

In 1911 a young man, named Niels Bohr, journeyed from his native Denmark to England to study physics. The following year he joined the great Rutherford in Manchester and began to work on the problem of the atom. Bohr was a keen supporter of the Rutherford solar system model of the atom despite its problematic nature.

Rutherford's model of the atom, in which the positive charge is concentrated in a nucleus whose radius is 10^{-13} cm around which electrons orbit with radii approximately 10^{-8} cm, was in deep conflict with classical electromagnetic theory on two accounts.

According to electromagnetic theory an electron orbiting a nucleus should emit a constant stream of continuous electromagnetic radiation and because of the ensuing loss of energy quickly spiral in to the nucleus. The apparent stability of the atom as well as the presence of discreet lines in the energy spectrum of atomic radiation had to be explained within the framework of Rutherford's model. It was this problem that Bohr set out to solve in 1912.

In addition to explaining the stability of the atom and the discreteness of its radiation spectrum, Bohr also had the task of explaining certain regularities of the spectral lines that had been observed by the spectroscopists. As early as 1885, Balmer had shown that the observed frequencies of the hydrogen atom could be represented by the following simple mathematical formula

$$f = \frac{R_y}{h} \quad \frac{1}{n^2} - \frac{1}{p^2}$$

Where f is the frequency of the radiation, R_y is Rydberg's constant equal to 3.27×10^{15} sec^{-1}, h is Planck's constant and n and p are integers almost always less than 10 or so. Balmer's formula provided a very tidy description of the spectral lines observed in 1885. As more and more data was collected it was found that the spectral lines of hydrogen occurred at precisely those places predicted by Balmer's formula. In 1908, Ritz discovered a regularity in the spectrum of light emitted by atoms, other than just hydrogen.

He found that if f_1 and f_2 are the frequencies emitted by an atom, then the same atom is very likely to emit the frequency $f_3 = f_1 + f_2$. This rule known as "Ritz's Combination Principle" follows trivially for hydrogen from Balmer's formulae. The novelty of his discovery, however, is that his principle applies to atoms other than hydrogen.

The work of Balmer and Ritz played an important role in Bohr's formulation of his atomic model by helping him make certain guesses he might not have made otherwise.

The quantization of energy proposed by Planck and Einstein played an equally important role in Bohr's thinking. He was the first to apply the quantum concept to the atom. He assumed that the electron orbits the nucleus only along certain fixed elliptical orbits, each of which represents a different quantum state.

He also assumed, in direct contradiction with electromagnetic theory, that the electron would not radiate while moving along any of the allowed orbits.

The electron is more or less stable in one of the allowed orbits or quantum states. Bohr did postulate, however, that the electron could jump to a lower energy orbit closer to the nucleus by radiating a quantum of energy. The electron would continue jumping to lower energy orbits emitting photons until it landed in the lowest energy orbit or ground state in which it could remain without ever emitting any more photons.

The frequency of the photon emitted by the electron as it jumps from an orbit with energy E_2 to an orbit with energy E_1 is related to the energy difference of the two orbits, $E_2 - E_1$. The relation between the frequency, f, of the emitted photon, the energy difference, $E_2 - E_1$ and Planck's constant, h, known as Bohr's frequency condition, is similar to Planck's original quantum condition, namely

$$hf = E_2 - E_1$$

According to Bohr's hypothesis, an electron not only may jump from an outer orbit to an inner orbit by emitting a quantum of energy, but it may also pass back from an inner orbit to a more energetic outer orbit by absorbing a quantum of energy. In order to make this transition it must absorb a quantum of energy exactly equal to the energy difference of the two orbits, $E_2 - E_1$.

Bohr's model of the atom violates the laws of classical mechanics and electromagnetism in a number of ways. First of all, the electron in the ground state is constantly undergoing acceleration without ever radiating light in direct contradiction with Maxwell's laws. Secondly, the frequency of radiation emitted by an electron as it jumps from one orbit to another is not equal to the frequency with which it orbits the nucleus. In classical electromagnetic theory, on the other hand, the frequency of the periodic motion and the frequency of the subsequent electromagnetic radiation are identical. Finally, in classical mechanics, an electron mat orbit the nucleus in an infinite number of paths, differing by only an infinitesimal amount of energy.

By placing this restriction on the angular momentum, Bohr was able to obtain Balmer's formulae for the radiated frequencies of the hydrogen atom. Bohr was also able to calculate Rydberg's constant, R_y and showed that it is simply related to the mass of the electron, m_e, the charge of the electron, e and Planck's constant, h, by the formula

$$R_y = \frac{2\pi^2 m_e e^4}{h^3}$$

This result, in which one of the fundamental constants of nature was related to the others, was a great success and insured the acceptance of Bohr's model.

This model not only explained Balmer's formula for the hydrogen atom but it also explained Ritz's combination principle. Let us label the quantum states or energy levels of the atom by E_1, E_2,..., E_n where E_1 is the energy of the ground state, E_2 is the energy of the first excited state,..., and E_n is the energy of the $n^{th}{}_{-1}$ excited state.

Here, we refer to the higher energy orbits of the electrons of the atom as excited states. These electrons have absorbed energy, but do not retain the additional energy very long. They shortly lose the excess energy by radiating one or more photons and finally drop back to the ground state. Bohr showed that the energy of the nth level, E_n, is equal to $-h/n^2 R_y$.

Let us consider the mth level whose energy E_m is greater than E_q but less than E_p.

Instead of dropping directly from level p to level q' by emission of a single photon, the electron in level p may first drop to level m through the emission of a photon with frequency f_{pm} equal to $(E_p - E_m)/h$ and then drop from level m to level q by the emission of a second photon of frequency f_{mq} equal to $(E_m - E_q)/h$. Adding the frequencies f_{pm} and f_{mq} we obtain f_{pq},

$$f_{pm} + f_{mq} = \frac{E_p - E_m}{h} = \frac{E_p - E_4}{h} + \frac{E_p - E_4}{h} = f_{pq}$$

Ritz's combination principle follows from the existence of atomic energy levels, the conservation of energy and Bohr's frequency condition.

Bohr's scheme correctly explained the spectroscopic rules of Balmer and Ritz. The model was severely limited, however, in the number of predictions it could make. For instance, there was no means of calculating the relative intensity of various spectral lines which experimentally differed from each other. Bohr's theory was also unable to predict the polarization of the light radiated by the atom. (The polarization indicates in which direction the oscillating electric field of the photon is aligned). Finally, not all the spectral lines indicated by the model actually occur experimentally.

Bohr's model was unable to predict which transitions were forbidden. Ironically, the classical theory, which is unable to explain the observed frequencies of the atom, is, however, unlike the Bohr theory, able to calculate the relative intensity of lines, the polarization and the forbidden transitions. In order to compensate for the deficiency of his model, Bohr incorporated the positive features of the classical theory into his scheme through the correspondence principle.

Bohr noticed that for the atomic transition among the more highly excited states that the difference in energy between the two adjacent levels becomes progressively smaller as the energy increases. He also noticed that the difference in the orbital radii also becomes smaller. As one goes to higher energies the transitions from adjacent levels becomes continuous as is the case in the classical theory.

The energy for all the atomic levels is negative. The reason for this is that the potential energy is negative and greater in magnitude than the kinetic energy ($1/2mv^2$) which, naturally is positive. As long as the electron is orbiting the atom and hence, bound to it, its total energy will be negative. If its total energy ever becomes positive then it will no longer be bound. When referring to high energy atomic level we will be discussing those levels for which n is large and, consequently, the energy E_n is almost zero but still negative. For these high energy levels, the difference between the nth level and the $(n-1)^{th}$ level is given approximately by

$$E_n - E_{n-1} \approx \frac{2R_y h}{n^3}$$

which goes to zero as n increases more rapidly than the energy E_n itself.

Bohr's theory of the atom was a hodge-podge of ideas. It incorporated the concept of the quantization of energy and violated a number of the basic rules of classical theory.

Through the correspondence principle, however, it still included the classical theory upon which it depended for its theory of polarization, the relative intensity of spectral lines and forbidden transitions. Despite its hodge podge nature, the Bohr model of the atom explained a surprisingly large number of the features of the spectroscopic data. Perhaps the most important prediction of the theory was the existence of discreet atomic energy levels. This aspect of the theory was dramatically confirmed one year after its formulation by Franck and Hertz.

Franck and Hertz studied the collisions of free electrons with the atoms of a gas, initially in an unexcited state. A beam of electrons with a fixed kinetic energy was directed at the gas atoms. The kinetic energy of the electrons after the collision was measured. As a result of the collision the free electrons transferred energy to the atom.

The transfer of energy from the electron can take place either by exciting the atom through an inelastic collision or by elastic scattering in which the atom remains in its ground state. Let us first consider the elastic case in which the atom is not excited. The atom will not gain very much velocity as a result of the elastic collision because its mass is so much greater than that of the electron. As a result, only an exceedingly small amount of kinetic energy is transferred from the electron to the atom and hence, the electron's energy loss is practically negligible.

The energy loss the electron suffers in the inelastic collision, on the other hand, is considerable. In this case, the atom is excited from its ground state to one of its excited levels. Because the energy levels of the atom are discreet, that is discontinuous, there is a minimum amount of energy which must be transferred to the atom before it can be raised to one of its excited levels.

This minimum amount of energy is exactly equal to the energy difference between the atom's ground state and its first excited state, E_2 - E1. If the kinetic energy of the incoming electron is less than this amount, then there is no possibility of an inelastic collision since the electron does not have enough energy to raise the atom to one of its excited states. This is exactly what Franck and Hertz observed.

As long as the kinetic energy of the electron was less than the threshold for inelastic collision, $E_2 - E_1$, they only observed elastic collision in accordance with the predictions of Bohr's theory. As soon as this threshold was exceeded, inelastic collisions were observed in which the electron's energy loss was equal exactly to $E_2 - E_1$. These collisions obviously corresponded to the excitation of the atom from its ground state to its first excited state.

As the kinetic energy of the electron is increased even further beyond the threshold, $E_2 - E_1$, other inelastic collisions are observed in which the atom is excited to even higher excited states. From the observed energy losses of the electrons, Franck and Hertz were able to determine the energy differences of the various atomic levels. These values were compared with the values obtained from spectroscopic data. The two sets of values for the energy levels were in complete agreement with each other. Franck and Hertz had verified Bohr's theory of the atom mechanically.

Another experiment which demonstrated the existence of Bohr's energy levels was devised by Maurice de Broglie, M. de Broglie bombarded the atom with x-rays of a known energy. He then observed the kinetic energy of those electrons ejected from atom as a result of absorbing the x-rays. For the difference of the photon's energy and the electron's kinetic energy M. de Broglie was able to determine the energy levels of the atom, which were also in agreement with those obtained from Bohr's theory.

In 1913, Moseley, in England, investigated the production of x-rays. His work revealed that the charge of the nucleus increased from one element to another by one unit of charge e. The relation of the energy of the x-rays emitted by an atom and the charge of its nucleus was found to be exactly that predicted by Bohr's theory of the atom.

The experiments of Moseley, M. de Broglie, Franck and Hertz established, beyond a shadow of a doubt, the validity of the basic concepts of Bohr's model of the atom such as the existence of energy levels and the Bohr frequency condition. As more and more experimental information was gathered, however, it became evident that Bohr's theory was not sophisticated enough to explain all the data.

When the spectroscopists looked closely at the spectral lines of Balmer, they discovered that each line was actually split into a number of finer lines. This fine structure of the spectral lines was explained by Sommerfeld making use of relativity theory.

The existence of the fine structure of the spectral lines revealed that, for each of the atomic orbits of a given radius postulated by Bohr, there are actually several orbits each with the same radius and almost the same energy but different ellipticities. The slight energy differences of these orbits arises from relativistic effects and accounts for the fine structure of each line.

The orbits postulated originally by Bohr are actually circular. Each orbit increases in size and has one more unit of angular momentum than the previous one.

Thus, for example, for the atomic level corresponding to the energy E_n, the radius of the orbit, R_n, is equal to $\frac{(n^2)h^2}{4\pi^2 me^2}$ and the electron in the orbit has an angular momentum equal to (n-1). (The ground state, E_1, corresponds to R = h^2/4? $^2me^2$ and zero angular momentum.). Sommerfeld discovered that, in addition to the circular orbits with energy E_n and Radius R_n, there are also elliptical orbits whose semi-major axis is exactly equal to R_n and whose energy is almost equal to E_n.. The angular momentum of these elliptical orbits, J, varies from J = 0 to J = (n-1). The ellipse with angular momentum, J = (n - 1), corresponds to the circular orbit postulated by Bohr. All or the orbits with radius R_n are more or less embedded in each other.

The very small variation in energy of the embedded elliptical orbits is due to a very complicated effect. The electron orbits the atom at velocities approaching the velocity of light and hence, relativistic effects must be taken into account. At these speed, the mass of the electron depends on its velocity.

The velocity of the electron varies along its elliptical orbit obtaining its greatest value at the point of its closest approach to the nucleus. The variation of the velocity and hence, mass is greater, the greater the ellipticity of the orbit. This accounts for the variation in energy of the various elliptical orbits corresponding approximately to the energy E_n. The variation in energy among the orbits with the same radius R_n is very small compared with the energy difference of two orbits with different radii such as R_n and R_{n-1}.

As a result of Sommerfeld's work, the orbits of the electron were referred to by two quantum numbers. One of the numbers, n, is the number used originally by Bohr to refer to the energy of the orbit. The other number is e and refers to the orbital angular momentum of each state. For each energy E_n, there are n different orbits corresponding to the values of the orbital momentum. The higher the orbital angular momentum of an orbit for a fixed value of n, the more circular is the path of its orbit.

In addition to the fine structure explained by Sommerfeld, it was also observed that the spectral lines split even more if the radiating atom is subject to an external magnetic field. This effect, known as the Zeeman effect, can be explained in terms of classical theory when a spectral line splits into three separate lines. The spectroscopists observed, however, that frequently a spectral line under the influence of a magnetic field split into more than three separate lines.

The splitting of the spectral lines is due to the fact that the electron orbiting the atom in an external magnetic field has additional energy by virtue of the existence of the external magnetic field. The amount of extra energy the electron acquires as a result of its interaction with the magnetic field depends on the value of its angular momentum, the orientation of the plane of its orbit with respect to the magnetic field, as well as the duration in which the electron rotates, i.e., clockwise or counter-clockwise. If the electron rotates in the counter-clockwise direction with respect to the external magnetic field, then the magnetic energy is positive and if the electron rotates in the opposite direction, the magnetic energy is negative.

If the magnetic field is perpendicular to the plane of the orbit then the magnitude of the magnetic energy is a maximum. If the magnetic field lies in the plane of the electron orbit, however, the magnetic energy is zero.

If the angle the magnetic field makes with the plane of the orbit, on the other hand, is between 0 degrees and 90 degrees then the value of the magnetic energy is between zero and the maximum (or minimum if the electron is orbiting in the opposite direction.)

The splitting of the spectral line for a given value of n and ? is clearly due to the different possible values of the magnetic energy which depends on the different possible values of the angle between the magnetic field and the plane of the orbit. Since all the values of the angle between 0 and 90 degrees would seem like possible orientation of the magnetic field with respect to the plane of the electron's orbit, one would expect all the possible values of the magnetic energy between the maximum and the minimum.

This is not the case, however, when one observes a finite splitting of the spectral line which corresponds to different levels. This result is completely baffling from a classical point of view and represents a pure quantum mechanical effect. Apparently, there are only 2n + 1 different orientations that the plane of the electron's orbit can make with the external magnetic field. In a sense, space is also quantized. Once a direction in space is defined by the external magnetic field only certain orientations of the electron's orbit are allowed.

A third quantum number, labeled m, has to be introduced to differentiate the possible orbits of the electron. The quantum number m gives the projection of the angular momentum vector (the direction of the angular momentum vector is defined to be perpendicular to the plane of the orbit) in the direction of the external field. The magnetic energy of the electron is therefore proportional to the value of m.

With the quantum numbers n, j and m the spectroscopists were able to account for many of the observed spectral lines. But there were still a number of mysteries which persisted. Some states seem to have half-integer values of the angular momentum. In order to explain the presence of half integer angular momentum, Uhlenbeck and Gaudsmit proposed that the electron spins on its axis and carries a half unit of angular momentum, thus, there is an analogy between the electron orbiting the nucleus of an atom and the earth orbiting the sun.

The earth possess angular momentum both by virtue of its spinning on its axis. The electron also has two sources of angular momentum, one is its orbital angular momentum equal to some integer, and the other is its spin angular momentum. If the sense of rotation is the same for both of these motions, both clockwise for example, then the angular momenta add and the total angular momentum equals, if the rotations are opposite, then the contributions subtract. The hypothesis of the spinning electron explains a number of the features of spectroscopic data which had previously defied explanation.

Direct experimental evidence for this hypothesis, however, is provided by an experiment of stern and Gerlach. In this experiment, a neutral beam of unexcited hydrogen atoms passes through a uniform magnetic field. The electron of the hydrogen atom in the ground state has zero orbital angular momentum. If the electron has no spin, then one would expect the hydrogen atom to pass through the beam unaffected since its total angular momentum would be zero.

When the orbital angular momentum is equal to ??, we discovered there were 2s + 1 possible orientations of the field with respect to the magnetic field. If the angular momentum is 1/2 s, one then expects (2 × 1/2 + 1) or 2 possible orientations of the spin. These correspond to the electron spinning on its axis in the counter-clockwise and clockwise directions which we refer to as spin up and spin down, respectively. The magnetic field will deflect the atoms with spin up differently than the atoms with spin down. One, therefore, expects that the beam of hydrogen atoms will divide into two beams, one with spin up and one with spin down, if the electron is a spin 1/2s particle. This is precisely what Stern and Gerlach observed verifying both the existence of the electron's spin and the quantuization of space. The same experiment was repeated with other atoms with higher values of angular momentum. In each case, the atom with angular momentum j? split into 2j+1 beamlets.

With the introduction of spin, a fourth quantum number, m_s, had to be defined to indicate whether the spin of the electron is up ($m_s = +1/2$) or down ($m_s = -1/2$). An electron in an atom is therefore defined by n, s, m and ms.

All electrons are identical. There is no way of distinguishing one electron from another. If an electron from an atom was removed and replaced by another electron with the same quantum numbers the atom would be identical. The principle of indistinguishability applies to all the other elementary particles in addition to the electron. There is no way of distinguishing one proton from another or one neutron from another. Two photons with the same frequency are identical. Some atoms are composed of electrons, protons and neutrons two atoms whose electrons have exactly the same quantum number and whose nuclei are in the same state are completely identical. This explains why the frequencies emitted by all hydrogen atoms are the same.

In 1925, Pauli discovered on the basis of his study of the energy levels of various atoms that no atom could contain two electrons with the same quantum numbers. Thus, for example, two electrons in an atom with the same values of n, s and m would be obliged to have their spins aligned in opposite directions, i.e., m_s = +1/2 and -1/2. It would be impossible to introduce into this atom a third electron with the same value of n, s and m as the first two. The electron would be excluded by the Pauli exclusion principle. The Pauli exclusion principle also explains the regularities of the periodic table of chemical elements.

In 1871, the Russian chemist Mendeleeff proposed a classification scheme of the chemical elements. He found that if he ordered the chemical elements by their atomic mass or weight the elements with similar chemical properties recurred periodically at more or less regular intervals. He constructed a table of the elements by increasing the atomic weight such that elements with similar chemical properties appeared in the same column. He left certain entries blank which he predicted would be filled by elements that had not yet been discovered.

He also described the chemical properties these missing chemical elements would possess on the basis of the chemical properties of the other elements in its column. The discovery of these elements displaying the properties predicted by Mendeleeff dramatically demonstrated the validity of his scheme. The x-ray work of Mosely in which he determined the nuclear charge of each element helped to refine the Mendeleeff classification scheme. For example, it explained why Argon, with atomic weight 39.9 but atomic number 18 comes before Potassium with atomic weight 39.1 but atomic number 19. A modern form of Mendeleeff periodic table.

The chemical properties of an atom are determined by the behaviour of its electrons, principally the outer electron which interact electromagnetically with the outer electrons of other atoms to form chemical bonds. The electrons in the atom arrange themselves in shells. Each shell corresponds to a different value of the principle quantum number, n. All the electrons in a given shell have the same average radius, R_n. The electrons in an atom under normal conditions settle into the state of lowest energy, the ground state. The inner shells are therefore the first to be filled. If it were not for the Pauli exclusion principle, all of the electrons of an atom would be in the n = 1 shell. Because of the Pauli exclusion principle, however, only a certain number of electrons are allowed in each shell.

The valency of an element is the number of electrons it gives up (if the valency is positive) or gains (if the valency is negative) when it enters into chemical combinations. Those elements with positive valencies which tend to readily give up electrons easily in chemical reaction also tend to be good electrical conductors. The valence number corresponds exactly to the number of electrons in excess of or missing from a closed shell as a study of the last eight entries to the periodic table reveals. Lithium, beryllium, and boron have the valencies +1, +2 and +3, respectively, as well as 1, 2 and 3 extra electrons in addition to their closed shell of two n = 1 electrons. Nitrogen, oxygen, and flourine have the valencies -3, -2 and -1 respectively, and are missing 3, 2 and 1 electrons from the n = 2 shell. Neon, with an n = 2 closed shell, has a valency of zero.

This leaves carbon which is more complicated since carbon can be viewed as having four electrons in excess of its closed n = 1 shell or missing four electrons from its n = 2 shell. We would expect on the basis of this, therefore, that carbon would enter into complicated bonds involving valencies of either +4 or -4. This is indeed the case which explains the extremely complicated chemistry of the carbon atom and why it is able to form molecules with long chains.

The next eight entries of the periodic table after neon correspond to the filling of the n = 3 shell. The eight outer electrons of the n = 3 shell of argon, an inert gas, form a closed shell. The eight elements, sodium, magnesium, aluminum, silicon, phosphorous, sulphur, chlorine and argon follow exactly the same pattern of chemical properties as the previous eight entries, with the valencies +1, +2, +3, +-4, -3, -2, -1 and 0, respectively.

The next entry in the periodic table does not continue to fill the n = 3 shell by populating the s = 2 states as might be expected. Instead, the two n = 4, s = 0 are populated in the next two entries, potassium and calcium.

The Bohr theory of the atom explained a great deal regarding the spectrographic and chemical properties of the atom. It had serious shortcomings, however. In spite of the correspondence principle it was never really able to successfully calculate the probability for a particular atomic transition and hence, could not account for the relative intensity of spectral lines.

There were also other aspects of the spectroscopic data that could not be explained within the framework of Bohr's model. Another difficulty was the inability to predict the frequencies of complex atoms in which more than one atom appears. Although Bohr's model was to be replaced with the more sophisticated wave mechanics, many features of Bohr's theory would be retained in the new quantum theory. Bohr's theory had been invaluable; for many years it provided a framework for organizing the wealth of information gathered by the spectroscopists.

Chapter 11

The Quantization of Energy

The exploration of the atom at the turn of the century provided the physicists of the day with a never-ending stream of surprises and shocks as one new discovery followed another. The most disturbing of all the new results of this period, however, occurred as a result of Max Planck's study of the energy distribution of electromagnetic radiation emitted from the surface of a hot body.

According to Maxwell's theory, an electric charge, when accelerated, emits electromagnetic radiation. As a body is heated, the atoms composing it begin to oscillate rapidly which causes them to emit radiation. In order to study the distribution of energy with respect to the frequency of this thermally induced radiation, a black body radiation was constructed.

It consisted of a hollow shell of metal in which a small hole was inserted. Since it is difficult for light to enter the interior of the hollow shell, the radiation inside the shell is due almost entirely to the thermal emission of electromagnetic radiation from the inner walls of the cavity. Any light which does enter the cavity is quickly absorbed by the inner walls. The distribution of energy of the thermal radiation can be studied by observing the radiation emerging from the hole in the shell. The intensity of radiation is plotted versus the frequency for two different temperatures.

The total amount of energy radiated increases rapidly with the temperature. It is proportional to the fourth power of the temperature. (If E is the total energy, T, the temperature, and k, a constant, then $E = kT^4$). As the temperature increases, the distribution shifts to the higher frequencies. The frequency for which the intensity is a maximum is proportional to the temperature.

The experimentally observed energy distribution differed very radically from the theoretical prediction depicted by the dashed cur. It was believed that the higher frequencies would dominate the distribution since the probability of exciting a frequency was thought to be greater, the higher the frequency. This prediction was not fulfilled, however, and the intensity, instead of steadily increasing with frequency, reached a maximum value and then decreased to zero. This result was a total mystery to the physicists of the day. No one was able to explain why the intensity did not continue to increase with frequency.

Max Planck devoted a great deal of effort trying to derive the experimentally observed distribution from a strictly theoretical point of view. Not meeting with success, he attached the problem from a phenomenological point of view. Rather than trying to derive the correct formula from first principle, he first searched for an algebraic formula which would just describe the experimental data. Using trial and error and guided by previous attempts at a solution, Planck finally found a formula which represented the data. Working backwards from the answer, Planck then searched for a way to derive his formula. He discovered he could obtain his formula if he made the assumption that the radiation of frequency, f, was absorbed and emitted in bundles of energy equal to hf where h is a constant, called Planck constant, and f is an extremely small number. The bundles of energies are called quanta (quantum is the singular) or photons.

The implication of Planck's result is that the energy of light is not continuous but is packaged in discreet bundles of energy called quanta or photons. According to Planck's hypothesis, a frequency, f, can not be excited unless an amount of energy, hf, is provided to create a photon.

Planck's constant, h, is extremely small and, hence, the amount of energy to excite a particular frequency is not much, however, it takes more energy to create the higher frequency photons that the lower frequency photons, and, hence, the probability of exciting the higher frequencies is not always greater than that of exciting the lower frequencies. Also, there is a cutoff of frequencies in Planck's model so that the infinite number of high frequencies can no longer contribute to the black body radiation.

If the total amount of energy available for thermal radiation is E_o, then, the frequency f_o, such that $E_o = hf_o$ is the highest possible frequency that can contribute. There just would not be enough energy available to create a photon with frequency greater than f_o.

Planck's hypothesis that the energy of light is quantized explained the distribution of energy of black body radiation. Because Planck's constant is so small and, hence, each quantum or photon actually carries such a small amount of energy, Planck's hypothesis does not conflict with the experimentally observed continuity of light.

Light rays are composed of literally millions and millions of tiny bundles of energy or photons, each of which carry a minuscule amount of energy. The amount of energy carried by the most energetic photon conceivable, for example, a photon emitted by a radioactive nucleus whose frequency corresponds to the gamma ray range, is equal to the amount of kinetic energy a drop of water would acquire falling under the influence of the earth's gravity the distance of 10^{-6} cm, the height of 100 atoms.

Although Planck's quantization of the energy of light did not violate any of the empircally observed facts of nature, the idea of the quantum violated the notions of classical physics held by his contemporaries. According to their way of thinking, most quantities and, in particular, energy are continuous.

The first break they had known in this tradition of continuity was the discovery of the atom which revealed that matter was discontinuous. With the discovery of the electron, the discontinuity or quantization of charge was also revealed. Although these concepts encountered some initial resistance, they could be incorporated into the fabric of classical physics because these discontinuities could be associated with the existence of particles which were always regarded as discreet. The discontinuity of the energy of light or electromagnetic waves was inconceivable, however.

Energy, first of all, was always considered a continuous quantity, even for discreet particles. But what made Planck's proposal even more mysterious was that it was associated with light which, ever since the diffraction experiments of Young, was considered to be a wave and, hence, continuous. Planck realized that his quantum hypothesis was in contradiction with the classical physics of his day. Referring to his work, he remarked to his son in Berlin in 1900,

"Today, I have made a discovery as important as that of Newton."

The full implication of Planck's idea was not understood for five years until Einstein exploited the quantum hypothesis in order to explain a new experimental result know as the photoelectric effect.

The photoelectric effect was first discovered by Hertz as early as 1887. It is the effect which has since been exploited in a number of devices such as the "electric eye" and the photographer's light meter. The effect consists of the observation that, when light of a sufficiently high frequency falls upon a metallic surface, electrons (referred to as photoelectrons) are ejected from the metal. This emission of photoelectrons from the metal can be understood from a classical point of view. The light which falls upon the metal is absorbed by the electrons inside. After a sufficient amount of time passes, the electron absorbs enough energy to overcome the electromagnetic forces which contain it in the metal and it is then free to leave the metal as a photo-electron.

There are three aspects of this photo-electric effect, however, which can not be understood in terms of classical physics. The first is the threshold effect of frequency. Unless the frequency of the light is above the threshold frequency, no photo-electrons are ejected from the metal, no matter how high the intensity of the light. There is no reason why this effect should depend upon the frequency as the experimental data seems to indicate. The second peculiar aspect of the photo-electric effect is that photo-electrons appear the instant the metal is irradiated by light. The electrons are ejected with 10^{-8} seconds of the light striking the surface of the metal, independent of the intensity of the light, as long as the frequency is high enough.

This contradicts the classical picture of the electron absorbing energy from the light wave, since it is easy to show that 10^{-8} sec does not allow sufficient time for the electron to absorb enough energy from a wave to overcome the electromagnetic forces which contain it within the metal. The third peculiar aspect of the photo-electric effect was pointed out by Leonard in 1902.

He observed that, when the intensity of light was increased by moving the source closer to the metal, for example, the energy of the electrons ejected from the metal did not increase but rather more electrons appeared. If one increases the frequency of the light projected on the metal, however, then the energy of the ejected electrons will increase.

Einstein was able to explain all three of these mysterious effects in his brilliant paper of 1905, on the photo-electric effect for which he was awarded the Nobel Prize in 1921 (Einstein never received a Nobel Prize for his work on relativity). Einstein expanded Planck's quantum hypothesis by assuming that, not only the light induced by thermal radiation is quantized, but, that all electromagnetic radiation comes in bundles of energy or photons.

The energy of an individual photon composing a light ray equals hf where h is Planck's constant and f is the frequency of the radiation. According to Einstein's hypothesis, at certain times, light behaves like a beam of particles where each particle is a photon. But, how does this explain, one might ask, the mysterious aspects of the photo-electric effect? Well, in order for an electron to escape the metal, it must have a sufficient amount of energy to overcome the normal electromagnetic forces which keep it captive inside the metal, i.e., it must have enough energy to overcome the binding energy.

If the frequency of the light is too small, then the energy of its quanta or photons is less than the binding energy and the photon can not transfer enough energy to the electron for it to overcome the forces keeping it prisoner in the metal. If one increases the frequency, f, such that hf is greater than the binding energy, then, the photons can deliver enough energy to the electrons to allow them to escape. This explains the threshold effect.

It also explains the instantaneous appearance of the electrons once the threshold of frequency is surpassed. The electron does not absorb the required energy for ejection from the metal accumatively from a wave, but, rather, all at once from a single photon.

The electron is ejected because it suffers a collision with a particle of light. Therefore, as soon as the collision occurs, the escape of a photo-electron is possible which is why electrons appear immediately the metal has been irradiated by a ray of light which is nothing more than a beam of photons.

The quantum hypothesis also explains why increasing the intensity does not increase the energy of an individual photo-electron. The energy of an ejected electron depends only on the frequency of the single photon which knocks it out of the metal.

Increasing the intensity of the light does not change the frequency of the photons, it only increases the number of them. This is why increasing the intensity increases the number of photo-electrons without increasing the energy of individual photo-electrons. The fact that the energy of the photo-electron depends on the frequency of the photon explains why the energy of the ejected electrons increase as the frequency increases.

Thus, Einstein was able to account for all the observationally-known facts concerning the photo-electric effect by assuming the particle-like behaviour of light. Not only did he explain all of the observations known at the time he wrote his paper in 1905, but he also made exact mathematical predictions relating the energy of the ejected photo-electrons to the frequency of the light inducing the effect. Let W represent the binding energy and f_0 the threshold frequency at which the photo-electric effects first occur.

Then the energy of the photons of frequency f_0 is hf_0 and equals the binding energy W. (In terms of an equation, we have $W = hf_0$ or $f_0 = W/h$). The energy of the escaped electrons ejected by photons with frequency f greater than f_0 is equal to the energy hf imparted by the photon minus the binding energy, W. (Defining E as the energy of the ejected electron and using an equation once again, we have $E = hf - W$).

This precise mathematical prediction, made by Einstein in 1905, was not verified until eleven years later in a series of experiments performed by R.A. Milikan (the man who measured the absolute charge of an electron using oil drops). He verified that the Einstein formula was correct and that the energy of the photo-electrons was proportional to the frequency.

The constant of proportionality is just Planck's constant, h, which Milikan found from his experimental results had the same value that Planck used when he explained black body radiation. Milikan's results confirmed, with great mathematical precision, that the energy of light is quantized and that light displays particle-like behaviour in the photoelectric effect. Another experiment which followed Milikan's photo-electric work also revealed the particle-like behaviour of light. This was the experiment of Compton in 1922 in which he scattered x-rays using electrons as the target.

When a beam of x-rays is directed at an amorphous (non-crystalline) solid-like graphite, most of the x-rays pass through the solid unscattered. Some of the x-rays are scattered in all directions by collisions with the electrons in the solid. Compton noted an effect which bears his name, namely those x-rays which are scattered by the electrons have a slightly smaller frequency after the collision and that the greater the angle of scatter, the more the diminution of the frequency.

The Compton effect is easily understood if the beam of x-rays is treated like a beam of photons. Each photon carries energy and momentum related to its frequency, f. The energy of the photon is equal to hf and its momentum to hf/c where c is the velocity of light. The momentum of a photon is related to energy by the theory of relativity and is equal to the energy divided by c.

Although the electrons in the graphite are bound to the nucleus, they may be treated as unbound because the energy of binding is so small compared to the energy of the x-ray photons. In analyzing his experiment, Compton treated the electron and photon as two particles colliding with each other. Using this assumption, he was able to explain all of the details of the Compton effect. Once again, electromagnetic waves behave as though they are a beam of particles.

Newton's corpuscular theory of light seems to be valid in certain circumstances, namely for black body radiation, the photo-electric effect and the Compton effect. For other situations, however, such as diffraction, interference and refraction, light behaves as a wave.

The dual nature of light provided physicists with the deepest mystery they had yet experienced. Light seemed to possess contradictory aspects. The dilemma was expressed succinctly by William Bragg who had worked on the x-ray diffraction experiments which show x-rays behave like waves. On Mondays, Wednesdays, and Fridays, the physicist believes and acts on the basis of wave theory; on Tuesdays, Thursdays and Saturdays, he follows the particle theory of light. To which some pundit must have added, "And on Sundays, he just prays for a resolution."

The prayed-for resolution was the modern theory of quantum mechanics that did not evolve until a number of equally nasty paradox were unearthed. The long and tortuous road to that resolution of the particle-wave duality of light began with the work of Niels Bohr.

Chapter 12

Dimensions of Electrical Quantities

QUANTUM ELECTRODYNAMICS AND ANTI MATTER

The quantum mechanics developed by Schrodinger and Heisenberg did not take into account the effects of Einstein's theory of relativity. The neglect of relativity did not limit the usefulness of the theory in any serious way, however. The velocity of the electron in an atom or molecule is never more that a few per cent of the velocity of light and, therefore, any relativistic corrections that are necessary are extremely small. Non-relativistic quantum mechanics proved to be an extremely useful tool in understanding the behaviour of atoms and molecules and hence, in understanding the behaviour of macroscopic materials such as solids, liquids and gases.

In spite of its many practical applications physicists were not completely satisfied with non-relativistic quantum mechanics. First, there were some small details which the theory did not completely explain. But perhaps more importantly there were aesthetic reasons for wanting to combine relativity and quantum mechanics.

Schrodinger, himself, as previously related had originally attempted to incorporate relativity into his wave equations but failed because of the complexity of the fully relativistic equations. Shortly after reporting his non-relativistic results, Klein and Gordon resurrected Schrodinger's original relativistic equation.

They were unable to do very much with it except to write it down and point out the difficulties that attended its use.

Paul Dirac also attacked this problem and was able to develop a different equation which was fully relativistic and avoided some of the difficulties of the Klein-Gordon equation.

Dirac found an equation to describe the electron which combined the results of both quantum mechanics and relativity. In addition, Dirac's equation also explained the spin of the electron and the presence of its magnetic properties.

Before Dirac's theory, it was impossible to explain the magnitude of the electron's magnetic momentum in terms of its spin. Any mechanical model of the electron which attributed its magnetic moment to the rotation of the electron's charge always involved velocities greater than the velocity of light and hence, contradicted relativity.

Dirac's equation, however, successfully related the magnitude of the magnetic moment of a charged particle to its spin, its charge and its mass. In addition to providing explanations for the spin and magnetic momentum of the electron, Dirac's equation was also able to explain the minute splitting of certain spectral lines which the non-relativistic quantum mechanics had not been able to explain.

Dirac's equation was a great success. It contained one difficulty, however, the resolution of which would prove to be interesting. Dirac's equation contained two types of solution. One solution corresponded to normal electrons with positive energies greater than or equal to m_ec^2, the rest mass energy of the electron. Another set of solutions corresponded to electrons with negative energies less than or equal to $-m_ec^2$. The existence of the negative energy states seem quite unfeasible experimentally.

If such states existed, electrons would spontaneously make quantum jumps from the positive energy states to the negative energy states emitting photons as they made this. Dirac excluded this possibility by a very imaginative and speculative application of the Pauli exclusion principle.

He postulated that all of the negative energy electron states were occupied and that there existed an infinite sea of negative energy electrons. Positive energy states because the Pauli exclusion principle does not allow an electron to make a transition into a state which is already occupied. This resolved the problem of the transition to negative energy states but then we are left with the problem of an infinite sea of negative energy electrons.

Dirac concluded that the sea of negative energy states is unobservable and corresponds to the vacuum state. The vacuum state is the state of no observable electrons. Although Dirac believed that the negative energy sea was not observable he assumed that one could observe the absence of an electron in one of the negative energy states.

A hole in the negative energy sea should behave like a positively charged particle since negatively charged particles would be attracted towards the hole to fill it and hence, would seem to be attracted by a positive charge. An analogous situation is observed in the behaviour of the Chlorine atom which is missing one electron to fill its outer shell. The missing electron in the Chlorine's shell acts like a positive charge since this gap attracts electrons to it and explains the intense chemical activity of Chlorine.

Dirac initially identified the holes in the negative energy sea with the proton. He thought that he had shown the mass of the hole was approximately 2000 times the mass of the electron. Pauli quickly pointed out that if the proton were a hole in the negative energy sea of electrons that the electron in the hydrogen atom would quickly jump into it and annihilate the hydrogen atom leaving only two photons behind to carry away the energy.

The mystery of the identification of he hole in the negative energy sea of electrons was resolved experimentally a short time later by the discovery by Anderson of particles which had the same mass as the electron but opposite charge. The new particles, called positrons, or anti-electrons, were detected in an experiment in which cosmic rays were observed by a cloud chamber.

Cosmic rays are high energy particles, produced in outer space, which enter the earth's atmosphere. The cloud chamber is a device used to detect the charged particles which leave tracks of ionization as they pass through the vapour of the chamber.

By immersing the entire cloud chamber in a magnetic field and measuring the amount and direction of the curvature of the charged particle's tracks, one can determine the mass and the charge of the particle from one's knowledge of the magnetic force. Anderson observed tracks in his cloud chamber with equal but opposite curvature of the electron. These tracks could only be made by a particle the same mass of the electron but with the opposite charge. The positron was quickly identified with a hole in Dirac's negative energy sea of electrons.

If Dirac's theory about the negative energy sea of electrons was correct, then one would expect that the electrons should jump into the holes in the negative energy sea and emit radiation. In this process, both the electron and the positron would be annihilated since the electron disappears by jumping into the hole and the positron or the hole disappears by being filled.

Once a positron encounters an electron the two particles annihilate each other leaving two quantas of light energy. The electron has jumped into the empty negative energy state releasing the energy gained by this transition as light energy.

The reason two photons are created is so that momentum will be conserved. This explains why the two photons go off in opposite directions. The amount of energy created by the annihilation is just equal to the total kinetic energy of the electron-positron pair plus the rest mass energy of the pair which equals $2mc^2$. Electron-positron pair annihilation not only confirmed Dirac's theory but also demonstrated the validity of Einstein's relation between mass and energy,

$$E = mc^2.$$

Dirac's theory of the electron also explains the creation of an electron-positron pair from light energy. When high energy photons pass close to a nucleus they are frequently converted into electron-positron pairs. In the cloud chamber this observation appears as the spontaneous appearance of a pair of tracks with equal and opposite curvature.

The photon does not leave a track because it carries no charge. From the initial direction of the electron and positron tracks, however, it is possible to determine the direction from which the photon came, and hence, the source which produced the photon. The energy of the photon which creates the electron-positron pair must be at least $2m_ec^2$ in order to provide enough energy to produce the rest mass for the two particles. The presence of the nucleus is only necessary to conserve momentum and energy.

The process of pair creation can be envisaged using Dirac's image of the negative energy sea. One may consider the photon as providing energy to an electron in one of the negative energy states and lifting it to a positive energy state. Since the electron in the negative energy sea is unobservable, it suddenly becomes visible when it is elevated to a positive energy state. The hole in the negative energy sea of electrons created by the absence of this electron also becomes visible as a positron. Pair creation is just the opposite process of annihilation. In pair annihilation an electron falls from positive energy to negative energy radiating light whereas in pair creation an electron absorbs light and jumps from a state of negative energy to one of positive energy.

The key concept Dirac used in interpreting his equation was the idea of the unobservable negative energy sea of electrons. Although this notion was essential in Dirac's development of his ideas, it is not absolutely necessary to the formulation of Dirac's theory. Having once gained an insight into how the electron behaves it is now possible to discard the concept of the negative energy sea and to view the Dirac equation as describing a charged spin one half particle and its oppositely charged spin one half anti-particle for which the processes of annihilation and creation occur.

The solutions to the Dirac equation which seem to correspond to negative energy can now be viewed as positive energy solutions of the anti-electron. The concept of the negative energy sea of electrons can be retained, however, to aid the mind's eye in conceptualizing pair creation and annihilation. If the idea of the negative energy sea were absolutely necessary to the formulation of Dirac's ideas, we would then have to deal with the mystery of how an infinite sea of negative energy electrons can remain unobservable. The sea of negative energy electrons does not exist in real space, it is a concept in the minds of those physicists who use Dirac's equation. Hopefully, it has also become a concept in your mind as well.

In our description of the annihilation of an electron-positron pair we failed to mention the very curious intermediate state that frequency occurs just prior to the annihilation process. The electron and positron on first encountering each other interact through their electric charges and often form a found state before annihilating analogous to the hydrogen atom.

The positron plays the role of the proton in the formation of this peculiar atom which is referred to by the name of positronium. All of the energy levels of positronium correspond to those of the hydrogen atom. All of the levels are shifted naturally because of the difference in mass between the proton and the positron. The entire Balmer series of frequencies for positronium have been observed and measured.

Once the positronium is formed, it makes radiative transitions from level to level until it reaches its ground state. It does not remain in this ground state for very long.

Because of the close proximation of the electron and the positron in the ground state, they eventually make contact and annihilate each other. The annihilation of the pair can take place from any of the levels of positronium. The probability is greatest, however, in the ground state because there is the greatest overlap the electron and positron wave functions in this state. The probability for annihilation is just equal to the overlap of the two wave functions, i.e., the probability of finding the electron and the positron at the same place.

Dirac's relativistic equation not only describes the electron but all spin 1/2 particles and hence, describes the proton and the neutron. We shall learn more about these particles when we study nuclear physics. For the moment, however, we shall introduce them into our discussion since Dirac's theory predicted that these particles would also have antiparticles, namely the antiproton and the antineutron.

The creation of a proton-anti-proton pair requires approximately 200 times the energy (the ratio of the proton mass to the electron mass) needed for the creation of an electron-positron pair. The observation of the anti-proton did not take place immediately following Dirac's prediction because it was impossible to create photons with the necessary energy, $2m_pc^2$, where m_p is the mass of the proton.

The discovery of the anti-proton had to await the development of particle accelerators which could yield such high energies. Such a machine was built in the early fifties at the University of California at Berkeley. Shortly after the completion of this machine the long awaited for anti-protons appeared exactly ad Dirac's theory had predicted. These particles have the same mass and spin as the proton but opposite charge. Anti-protons and protons annihilate each other to produce photons in the same way as electrons and positrons. The reaction of the proton and anti-proton is more complicated, however, because of the nuclear force but the essential features of Dirac's theory still persist. Shortly after the discovery of the anti-proton the anti-neutron was also detected.

This particle has the smae mass and spin as the neutron. Its charge is also equal to zero since the opposite of zero is still zero, i.e., (-0 = 0). Neutron anti-neutron pairs are created and annihilated just as expected by Dirac's theory. One of the interesting features of the antiparticles is that their interactions among themselves is exactly identical to the interactions of the particles among themselves.

The anti-proton anti-proton interaction is the same as the proton-proton interaction. A positron and an anti-proton interact with each other just like the proton and an electron and hence, form an anti-hydrogen atom which behaves exactly like the hydrogen atom.

The anti-hydrogen atom has been observed and found to emit the same spectral lines as the hydrogen atom. There is no way of distinguishing matter from anti matter. Given enough positrons, anti-protons and anti-neutrons, one could build a world identical to our own. In fact, it is purely convention to call the material of which we are made matter and those particles which annihilate our particles anti-matter.

For all we know, we are made of anti-matter and the other particles are matter. There is no way of determining whether the other galaxies or clusters of galaxies in our universe are made of matter or anti-matter. There are, in fact, cosmological models which predict that half the galaxies in the universe are made of anti-matter and the other half of matter. We shall explore these ideas later in the book.

In addition to the relativistic equation Dirac developed for the electron, he also developed an equation to describe the photon and its interaction with electrons and other charged particles. Dirac's equations gave results in excellent agreement with experiment.

They described all of the basic interactions of electrons and photons and included the processes of pair creation, pair annihilation, the emission of radiation and the scattering of electrons and light. As experimental measurements of energy levels of atoms improved, it was discovered that there were extremely small discrepancies between Dirac's theory and experiment. Through the work of Tomonaga, Schwinger and Feynman in 1948, it was discovered that these discrepancies could be accounted for by calculating corrections to Dirac's theory.

The quantum electrodynamics developed by these physicists and their co-workers yielded results of unprecedented accuracy. They were able to calculate corrections to the energy levels of the hydrogen atom to one part in 106. They were able to explain the 1% discrepancy between the experimental value of the electron's magnetic moment and the one predicted by Dirac theory. They predicted that the magnetic moment of the electron would be equal to

$$\mu THEO = \frac{eh}{4\pi m_e e}(1.001159655 \pm 3)$$

whereas the measured value is

$$\mu EXP = \frac{eh}{4\pi m_e e}(1.001159657 \pm 4)$$

The two numbers agree to 1 part in 109. The calculation of the electron's magnetic moment by quantum electrodynamics represents the most accurate determination of a number in atomic physics.

We could never come to understand the intricacy of quantum electrodynamics without delving into the mathematics of the theory. We can, however, gain an understanding of the physics behind the theory and at the same time gain a deeper understanding of the nature of the static electric force.

Perhaps the most mysterious aspect of the electric force is the concept of action at a distance. It is very hard to conceive how two charged particles separated by a finite distance can interact with each other with no medium between them.

A hint of how this can take place, however, is provided by the production and detection of light. Let us consider the process whereby we see the light produced by a lamp. Light is produced within the filament of the light bulb by exciting the atoms of the filament.

The atoms are excited by collisions with the electrons passing through the filament as an electric current. The photons are actually emitted by electrons jumping from one orbit to another within their respective atoms. The photons or light travels through empty space as small packets or bundles of energy.

The stream of photons enter our eyes and are absorbed by the electrons in our retina. As the electrons absorb photons, they also absorb energy which causes them to move about. The motion of these electrons is converted by the optical nerve into an electrical impulse which is transmitted to our brain where we become aware of the light.

What can we learn from this description of the production and the detection of light?

What I find most interesting is that the moving electrons in the light filament caused electrons in the retina to move also through the exchange of light or photons between them. The photons were the medium that allowed the electrons in the filament to move the electrons in the retina.

In a sense, the photons transmitted a force from the filament electrons to the electrons of the retina.

Perhaps we can apply this lesson to help explain how the electric force between static charges is transmitted. The principle of the conservation of energy makes direct application of the analogy possible. One can not simply claim that the electric force arises from the exchange of photons because one must account for the energy necessary to produce the light. In the case of the electrons that are moving the energy necessary to create the photons came from the de-excitation of the atoms of the filament and the energy that was absorbed by the retinal electrons went into the kinetic energy of their motion.

In the case of static charges, however, the particles remain at rest before and after the transmission of the force. The electrostatic force, however, is independent of the motion of the charges. If the charges remain static throughout the electrostatic interaction there is no way to provide the energy for any photon that may be transmitted between the two charges.

This is best seen by considering the time just prior to the hypothetical exchange of a photon, the time during the exchange and the time just after the exchange.

Just before the exchange and just after the exchange the total energy of the system is just equal to the energy of the two static electrons. But during the time of the transmission of the photon there is, in addition to the energy of the two static electrons, the energy of the exchanged photon, and hence, during this time there is a violation of energy conservation.

This would seem to kill our hypothesis. However, if we involke the uncertainty principle it is possible to resurrect our scheme.

A violation of the conservation of energy is meaningless if it can not be measured. If there is no way of detecting the photon during the time of its exchange then we do not have to worry about the conservation of energy since the total energy before and after the exchange remains the same. We, therefore, shall assume that the static electric force arises from the exchange of a "virtual photon" between two charged particles. We call the photon a "virtual photon" because it can not be detected directly.

Let us investigate whether the exchange of a virtual photon is inconsistent with conservation of energy and the uncertainty principle. Unfortunately, this investigation will require some rudimentary mathematics which should not disturb most readers.

We shall show that the uncertainty principle does not permit an observation of the exchanged photon and hence, the photon is virtual. We shall also show that the only form of the force consistent with the hypothesis of virtual photon exchange, the uncertainty principle and conservation of energy is the inverse square law. The uncertainty principle has two forms; one of which states that the product of this uncertainty in measuring the position and the momentum must be greater than h, Planck's constant (or $p \times h$).

The other form of the uncertainty principle states that the product of the uncertainties in measuring the lifetime and energy of a system is greater than h or ($E\ t\ h$). It is the latter form that we shall exploit now. The system of the virtual photon will have a lifetime,

$$t = R/c$$

since this is the time required for the photon to propogate the distance, R, between the two charges. We assume the energy of the virtual photon is related to the potential energy of the two charged particles which in turn depends on the force. The force, we learned, is inversely proportional to the distance between the two charges. If the two particles are electrons then the force is given by,

$$F = \frac{e^2}{R^2}$$

Aside from the numerical constants which in this case equals 1, the potential energy, E_p, is related to the force, F, by the expression

$$E_p = FR$$

and hence

$$E_p = \frac{e^2}{R}$$

The product of the energy of the photon times its lifetime is, therefore, just given by

$$Et = \frac{e^2}{R} \times \frac{R}{c} = \frac{e^2}{c}$$

The quantity e^2/c has the same dimensions as h and just happens to equal h/137. Since the product of the quantities that we wish to measure is 137 times the product of the uncertainty of measuring these quantities it is quite evident that it will be impossible to detect the virtual photon, and hence, our hypothesis of photon exchange is not inconsistent with the law of conservation of energy.

It is interesting to note that the distance R between the two charges dropped out of the expression for Et. This result only occurs for the inverse square law.

If the force had any other form the expression for E x t would have depended on R. It would then be possible to find a value of R for which Et became greater than h and hence observable in principle. If this were to occur then we would have to give up the notion of virtual photon exchange and look for another explanation of action at a distance. For example, if the force were e^2/R^3 then the potential energy would be $e^2/2R^2$ and the product of energy times lifetime would be

$$Et = \frac{e^2}{2R^2}\frac{R}{c} = \frac{e^2}{2c} = \frac{h}{274}\frac{1}{R}$$

By considering R small enough it would be easy to make Et greater than h and hence, the photon detectable.

The hypothesis of virtual photon exchange gives us some understanding of why the electric force goes like the inverse square of the distance. It is the only dependence of the distance R which is consistent with the uncertainty principle and the conservation of energy. The uncertainty principle plays an extremely interesting role in the virtual photon hypothesis. Rather than limiting our knowledge of the universe the uncertainty principle enables us to understand certain aspects of basic physics more deeply.

The hypothesis of virtual photon exchanges explains how action at a distance occurs. Electrons interact through the exchange of virtual photons. Individual electrons are continuously emitting and reabsorbing virtual photons. They are surrounded by a cloud of virtual photons. Whenever another charged particle enters that cloud it will interact electrically (i.e., experience a force) by absorbing one of the virtual photons. The electron whose virtual photon is absorbed will also experience a force because the absorbed virtual photon will never return to balance the momentum lost by its original emission.

Quantum electrodynamics is based on the hypothesis of virtual photon exchange. Dirac's theory is based essentially on the exchange of a single photon between charged particles. Quantum electrodynamics provides a more detailed description of charged particles by taking into account the exchange of more than one photon. Multiple photon exchange can be represented pictorially with diagrams first devised by Richard Feynman. The following figure shows the diagram for single photon exchange.

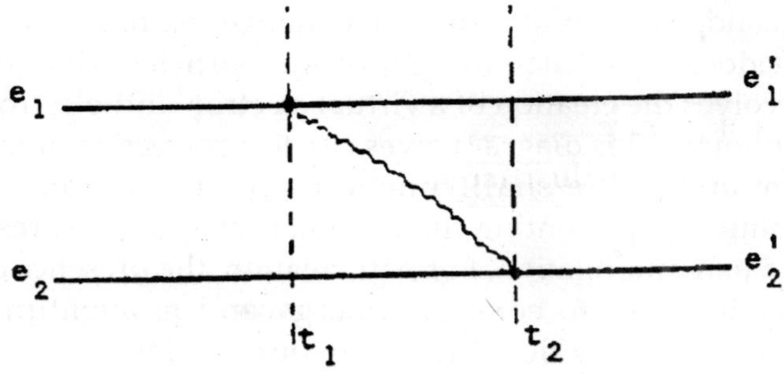

Time is increasing in this diagram from left to right. The lines labeled e_1, e_2, e'_1 and e'_2 represent electron 1 and 2 before and after their mutual interaction. At time t_1, e_1 emits a virtual photon represented by the wiggly line which is absorbed by e_2 at time t_2. Note that as a result of the emission at t_1, e_1 changes to e'_1 and as a result of the absorption at t_2, e_2 changes to e'_2.

The first two diagrams are equivalent to each other. The same is true of the last two so that there are only two types of diagrams involving the exchange of two photons. The second set of diagrams are corrections to the first because they make a much smaller numerical contribution than the one photon exchange.

Each time a photon is exchanged a factor of e^2/hc enters the probability of the exchange taking place. Since

$$e^2/hc = 1/137$$

the relative probability of two photon exchanges to one photon exchange is 1/137. The relative probability of n photon exchange to one photon exchange is $(1/137)^{n-1}$. Each successive order of photon exchange becomes less and less probable. Nevertheless, these small corrections have been calculated and have been shown to agree with experiment as pointed out above.

In addition to the diagrams in which two photons are exchanged between the charged particles, there are other two photon diagrams in which only one photon is exchanged. We give three examples of these types of diagrams:

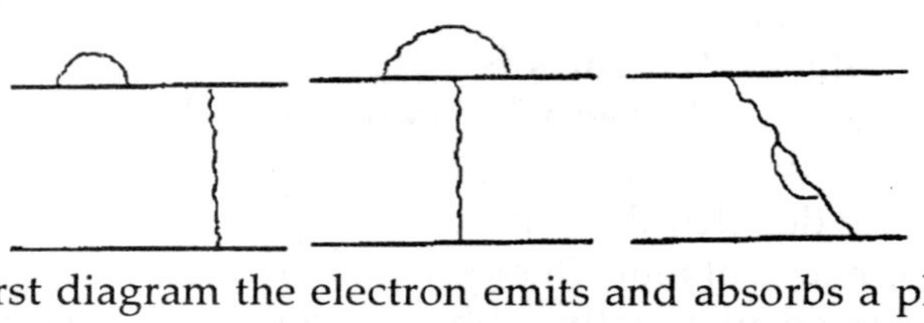

In the first diagram the electron emits and absorbs a photon before exchanging the photon with the other charged particle. This type of diagram does not change the nature of the electric interaction but it does not change the mass of the charged particle.

The second diagram also does not change the nature of the electric force but it does affect the strength of the particle's charge. The third diagram involves the creation of a virtual electron-anti-electron pair from the virtual photon. This diagram gives rise to a correction of the electron's magnetic moment. We shall conclude our discussion of quantum electrodynamics by presenting the Feynman diagrams representing pair creation and pair annihilation. For pair creation, the presence of a nucleus is necessary in order to conserve energy and momentum. We shall represent the nucleus by the symbol z in our diagram

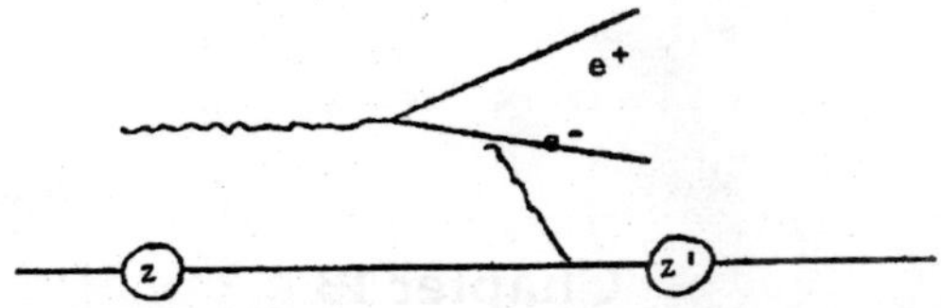

In the process of creation we see a real photon and a nucleus entering. The real photon turns into an electron positron pair. The electron then exchanges a virtual photon with the nucleus. Pair annihilation occurs in the following diagram through the spontaneous emission of a photon by the electron followed by the annihilation of the electron and positron into the second photon:

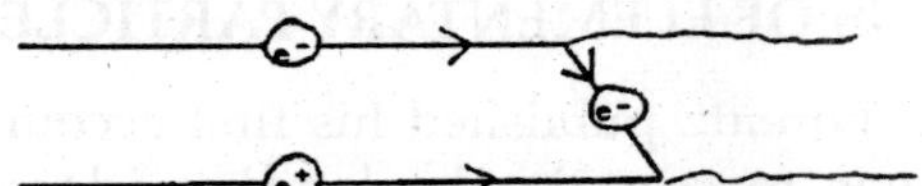

One can look on this diagram as the e$^-$ exchanging a virtual e$^-$ and turning into a photon. The positron then absorbs the virtual e- and also turns into a photon. The last diagram has an interesting interpretation if one assumes that a positron is an electron moving backwards in time. Then the pair annihilation has the form

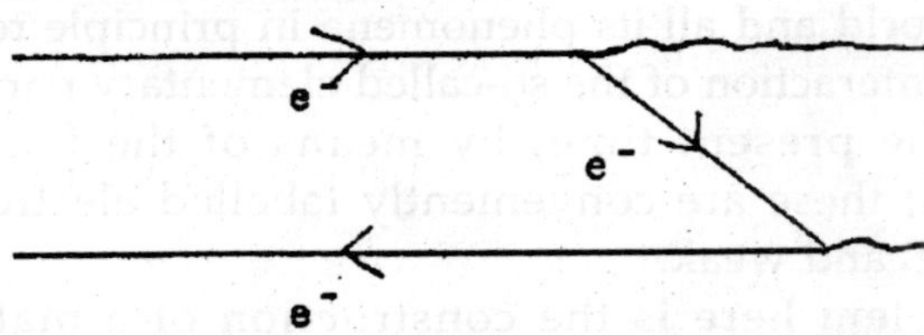

where it would appear the electron enters the diagram at the top and changes its direction in time and exits the diagram at the bottom.

Chapter 13

Electron Theory, Atom

THE BEGINNINGS OF ELEMENTARY PARTICLE THEORY

In 1892, H. A. Lorentz published his first account of the theory of electron which was to contain the leading ideas of many future investigations for more than a generation to come. It should be noted that this event was prior to the experimental clarification of the basic properties of the electron; those were not measured until 1897 and 1898 when J.J. Thomson succeeded in a determination of the charge to mass ratio and in a first measurement of the elementary charge *e*.

The physical world and all its phenomena in principle reducible to the problem of the interaction of the so-called elementary particles (over thirty known at the present time) by means of the four types of fundamental forces; these are conveniently labelled electromagnetic, gravitational, strong, and weak.

The basic problem here is the construction of a mathematical framework in which the static and the dynamical properties of these particles and their interactions can be reduced to a minimum amount of experimental information. If we know the mass of one particle, how can we compute the masses of all the other particles? Is there one interaction of which the electromagnetic and the gravitational interaction are special eases? The answer to questions of this type is the ultimate goal of physicists working in fundamental theory.

So far, we have barely scratched the surface concerning these matters. Of the four interactions, the electromagnetic one is certainly most thoroughly investigated and best understood. But it is not at all completely understood. Of the electrically charged particles we know most about the electron. The theory of the electron has throughout its history been very closely linked to the theory of fundamental electromagnetic interactions. It has been the model and guide to understand the other interactions.

THE AGE OF LORENTZ

The mathematical formulation of *macroscopic* electromagnetic phenomena was beautifully accomplished by J. C. Maxwell about thirty years prior to Lorentz's theory. Lorentz constructed a *microscopic* theory by using Maxwell's equations and adding to it an expression for the force which a charged particle experiences in the presence of electric and magnetic fields. This microscopic theory is a description of matter in terms of its charged atomic fragments, ions and electrons. The success of this microscopic theory lay in the proof first provided by Lorentz that the macroscopic Maxwell theory can be deduced from this microscopic theory by a suitable averaging process over the motion of the individual ions and electrons. Thus, Lorentz's theory became the primary theory and Maxwell's theory can be reduced to it.

However, Lorentz went beyond this: having successfully described the electromagnetic force acting on a charged particle due to externally present fields, he attempted to describe the structure of an individual electron. His aim was to show that the electron is a completely electromagnetic object. In particular, its mass was to be the mass equivalent of its electromagnetic energy contents; its inertia, i.e., the inertial term in Newton's equations of motion, was to be entirely due to its own electromagnetic field.

Accelerating the electron means changing or distorting the field produced by the electron; this requires work. Therefore, the electron exhibits a certain inertia in following the force acting on it. These ideas were contained, partly implicitly, in Lorentz's work. They were later clarified and extended to fast moving electrons by Abraham and others especially after Einstein's special theory of relativity became accepted. However, relativity was not applied to the theory in a consistent way and, consequently, the difficulty that the union of relativity and electron theory could have removed, remained there for many years. Let us therefore look at the nonrelativistic theory. As mentioned previously, the starting point of the theory is the Lorentz force and the microscopic equations which, when averaged, produce Maxwell's equations. When the Lorentz force is used to describe the action which the electron's own electromagnetic field exerts on its source, the electron, the following equation of motion is obtained:

$$\frac{4}{3}\dot{i}ra\frac{2v^3}{3c^3}\dot{a} + \text{"structureterms"} = F^{(1)}$$

In this equation the acceleration a and the force F are three-vectors and the dot indicates a time derivative. m and e are mass and charge of the electron and c is the velocity of light. The famous Lorentz force F is due to the external electric field E and magnetic flux density B, and is a sum over the charge density ρ,

$$F = \int \rho (E + v^k B) d^3 x.$$

The electron's structure is therefore characterized by a charge density distribution. This quantity has to be assumed, since there is nothing in the theory which would determine it. Only the total charge e is known from experiment.

The mass m is completely electromagnetic and is therefore given by the energy W of the electric field when the particle is at rest,

$$m = \frac{W}{c^2} \cong \frac{c^3}{r_0} \frac{1}{c^2}.$$

In this equation the radius $r0$ of the electron enters.

Equation (1) now exhibits three difficulties, associated, respectively, with each of the three terms appearing on the left hand side:

(I) The inertial term differs by a factor 4/3 from Newton's classical "mass times acceleration."

This is a kinematic problem which implies that the relationship between momentum and velocity for a particle in Newtonian mechanics differs from that for the completely electromagnetic electron. This would have dire consequences and is an intolerable defect of the theory. Luckily, it can be corrected rather easily. A conscientious merger of this theory with special relativity assures that this factor disappears, since it is incompatible with the relativistic transformation properties. For a finite electron this was first pointed out by Fermi in 1922. It is closely related to the definition of rigidity in special relativity where the difference in simultaneity of relatively moving observers plays an essential role. Unfortunately, Fermi's paper was either never understood or soon forgotten. In any case, the factor 4/3 can still be found in some of today's texts.

For point electrons, the removal of this factor was later rediscovered several times. Let me summarize then by saying that a relativistic generalization of (1) will not show this factor; if the non-relativistic theory is derived as the limit of the relativistic one, this factor will disappear from (1). A study of its origin reveals an unjustified and incorrect assumption about the relationship between the Poynting vector and the momentum of non-radiative electromagnetic fields.

(II) The second difficulty lies in the terms marked "structure terms." These terms depend on the charge distribution and radius of the electron. This dependence is a difficulty here while it is not a difficulty in the first term. In the latter, the mass must be determined experimentally anyway, while here the whole dynamics becomes explicitly dependent on the electron structure. This difficulty can be removed only by eliminating the electron structure altogether. If we assume the particle to be a point particle it will obviously have no structure. Furthermore, if, for this purpose, we let the electron radius *r*0 shrink to zero the "structure terms" all vanish.

Since no experimental evidence for structure exists (the Coulomb potential is correct down to smallest distances measured) this is a satisfactory procedure indeed. The removal of the structure terms, however, produces a new difficulty, viz., that when the radius shrinks to zero, the electron mass becomes infinitely large. According to (3),

$$r_0 \rightarrow 0 \text{ implies } m \rightarrow \infty$$

This is the famous problem of the electron self-energy. It exists in the classical theory of the electron as well as in the quantum theory. A satisfactory solution of it is not known. An at best temporary solution is provided by the renormalization procedure and will be discussed later. In any case, this procedure no longer allows for a completely electromagnetic mass.

It is therefore not surprising that many attempts were made to keep the electron of finite extension and carry along throughout the development of the theory an explicit electron structure. The debate pro and con the point electron is in fact as old as electron theory. Bothsides agree, however, that since the advent of quantum mechanics, the problem of the structure of the electron has become a quantum mechanical one. Surely, it does not belong into the field of competence of classical electrodynamics.

Consequently, it is to some extent a matter of taste, exactly how the classical theory describes the electron *outside* its limits of validity. The classical theory is not competent to discuss structure, no structure should be introduced if at all possible. In fact, the most physicists would agree that a classical point electron theory is preferable if it can be constructed without divergences. This cannot be done on the Lorentz level of development, but, as we shall see, it can be done at a later stage.

(III) The third difficulty evident in the Lorentz equation (1) is the following. Even if the structure terms were absent, there is still a term in the equation which involves a time derivative of the acceleration. Now Newtonian mechanics is based on equations of motion which involve no such time derivatives. Consequently, they are second order differential equations in position, and their solutions are determined by *two* initial conditions: if the position and the velocity of a particle is given at the starting time, its whole trajectory is completely determined. Not so for the Lorentz equation: there we have at least a third-order equation (viz. when there are no structure terms) and, therefore, the initial *acceleration* would have to be known too.

This completely contradicts the Newtonian concepts of mechanics. The term also has a rather nebulous physical interpretation. It corresponds to a reversible change of energy, somewhat like the kinetic energy. In fact, this energy was named "acceleration energy" in contradistinction to the "velocity energy" i.e., the kinetic energy. After Schott, it is sometimes called the Schott energy. While the second term in (1) obviously arises from the self-field of the electron and is therefore often referred to as "radiation reaction."

There is one more difficulty in the Lorentz-Abraham theory which is not apparent from the equation of motion. This difficulty relates to the stability of the electron. Obviously, an accumulation of negative charge is not a stable configuration: Any finite charge distribution of only one sign would explode.

The completely electromagnetic theory of the electron there cannot even account for the very *existence* of this particle. This is a very old problem.

A solution to it was proposed by Poincaré who showed that an attractive and consequently non electromagnetic force can always be added to the theory so as to just balance the stresses and establish stability. It is a very *ad hoc* solution and not at all fitting a "fundamental" theory.

On the other hand, in the point electron theory the electromagnetic self- stress is infinite and so must be the Poincarés cohesive force. This divergence is closely related to the self-energy divergence of the point electron. In fact, one can show that in a relativistic theory in which the electron self-energy (i.e., its mass) is not infinite, the self-stress will actually vanish and the particle will be stable.

Thus, this problem need not be further discussed since it will be resolved simultaneously with the self-energy problem in any relativistic theory. Physically, this is quite reasonable, because, if the electron is the elementary charge, a satisfactory theory must be so as to admit no Coulomb-type self-interactions, since such interactions exist only *between* charges.

Therefore, in such a theory it is no longer possible that *part* of the elementary charge acts on another *part* of the same elementary charge. Consequently, the electron will no longer explode, but will be stable.

THE SUCCESS OF THE CLASSICAL POINT ELECTRON THEORY

The next major development occurred in 1938 when Dirac published a most important paper. He started with the microscopic Maxwell-Lorentz equations. By a judicious use of the conservation laws and a covariant formalism throughout, he derived an equation of motion for a point electron which in the non-relativistic limit is exactly the Lorentz equation (1), but without the factor 4/3 and without the structure terms,

$$ma - \frac{2}{3}\frac{v^2}{c^3}\dot{a} - F - v.$$

Here, a, v, and F are four-vectors of acceleration, velocity and force. The new term on the right involves the radiation rate, $\mathfrak{R}$, i.e., the energy per second of electromagnetic radiation emitted at a given instant by the electron.

This term vanishes in the non-relativistic limit. Since this equation is almost exactly the Lorentz equation in the point particle limit, and in order to avoid confusion with Dirac's equation for the electron in relativistic quantum mechanics, this equation is referred to as the "Lorentz-Dirac Equation."

In fact, the radiation term was also known long before Dirac. The difference between (5) and Newton's ma = F first obtained by Abraham as the relativistic generalization of the Lorentz radiation reaction term.

These two terms are known as the "Abraham Radiation Reaction Four-vector." At this point one must note that, while the term has a rather questionable physical meaning, the $\mathfrak{R}$v term has a very well defined meaning. It is exactly the rate at which energy and momentum in the form of radiation leaves the electron.

One more comment is essential concerning the derivation of the Lorentz-Dirac equation. In addition to the Maxwell-Lorentz equations one new assumption and one new procedure enter this derivation. The assumption is that the equation be of third order at most and not higher.

As Dirac put it in his 1938 paper concerning higher order equations: "...they are all much more complicated,... so that one would hardly expect them to apply to a simple thing like an electron." The new procedure used is that of "mass renormalization." This means that the mass equivalent of the electromagnetic self-interaction is added to whatever other mass the electron has ("bare mass") and the sum is identified with the observed mass.

This originally classical procedure was not sufficiently appreciated at the time and was not used in the quantum domain until almost ten years later, yielding essential advances in quantum field theory; it became the basis for the successful computations of radiative corrections. The elimination of the kinematic difficulties in Dirac's approach is not surprising.

As mentioned earlier, it is assured by the consistent relativistic treatment used. Neither is it surprising that this point-electron equation contains no structure terms: there is of course an *infinite* electromagnetic mass term hidden in the renormalization procedure. For this reason the point electron limit can be carried out after renormalization. What is remarkable is that here Dirac "almost" succeeded in deriving the equation of motion of the electron from the

Maxwell-Lorentz equation. It is well-known that the Lorentz force is a separate assumption, independent of Maxwell's equations. But here it is almost reduced to them.

If this were actually accomplished, one would have a situation very similar to general relativity. The field equations there imply the equations of motion. That means that the equations which permit the computation of the gravitational field from the motion of the masses also determine the motion of these masses under the influence of the gravitational fields. The equations of motion are part of a self-consistency requirement for the field equations.

The main difficulty that still remains in the Lorentz-Dirac equation (ignoring for the time being the infinite mass renormalization) is the appearance of å term: the equation is of third order rather than of second order. This difficulty already exists in the Lorentz equation, but because of the structure terms it was perhaps less acute. The term requires the initial acceleration for a complete determination of the solution.

In order to appreciate the import of this situation, it is necessary to mention a difficulty which is not evident from the equation and which is actually closely related to the å term. If the initial acceleration is not specified, an infinity of solutions results, all which—with only one single exception—are physically meaningless. They lead to trajectories according to which a particle would accelerate and approach the velocity of light in the distant future, no matter what the applied force is. In fact, even when no force at all is acting on the particle, such solutions exist.

These solutions are therefore called "self-accelerating" or "run-away."

It follows that the initial acceleration is not only necessary to determine the solution uniquely, but the "right" initial acceleration must be specified in order to assure the *physical* solution.

The smallest error in the initial acceleration yields a run-away solution. How can these non-physical solutions be avoided? What principle or condition must be invoked in order to permit only the physical solution? Dirac mentioned that if one required that the acceleration should vanish in the distant future, one will have no run-away solutions. But this is almost a tautology. Furthermore, it is an *ad hoc* condition, since no physical principle seems to be involved.

However, a much better approach is possible. If one invokes the fact that only a finite amount of energy is available for a given system, then it follows from energy conservation that a physical system can radiate only a finite amount of energy, even if it is radiating for an infinite time. Thus, we require

$$\mathfrak{R} \to o \text{ for } \tau \to \infty .$$

Otherwise, the selection of the physical solution is not assured by the equation itself. The condition (6) is not an asymptotic condition imposed to specify a particular problem, as is done, e.g., in scattering theory, but it is a necessary condition which must be fulfilled no matter what specific system is under consideration.

By a few simple mathematical steps one can now prove the following. The Lorentz-Dirac equation and the asymptotic condition are together equivalent to the one equation where K stands for the right hand side. From what has been said, it follows that this equation should be taken as the equation of motion for the electron instead of the Lorentz-Dirac equation, since the latter is not meaningful without the condition.

The new equation of motion has many remarkable properties:

- It has no run-away solutions, since it was constructed that way.
- It is an integro-differential equation of second order, so that its solutions are determined in the Newtonian spirit of mechanics by initial position and velocity only. The transition eliminated the term å and the run-away solutions at the same time.
- The absence of the å avoids the question of that term's exact physical meaning.
- Only the term $\mathfrak{R}v$ occurs as radiation reaction. This term vanishes if and only if there is no radiation $(\mathfrak{R} = o)^{v}$ which was not the case previously. This has led to a great deal of confusion in connection with uniformly accelerated charges which seem to radiate with zero radiation reaction.
- The solutions of this equation are exactly consistent with an extension of the principle of equivalence to electromagnetic systems. In particular, a neutral and a charged particle in a static homogeneous gravitational field fall equally fast. Nevertheless, an observer supported in such a field will see the charged particle radiate. This leads to an apparent contradiction with energy conservation which is resolved by Equation as follows: When this equation is written

  $$ma(r) = F(r + \xi\, r_0) = \mathfrak{R}\,(r + \xi\, r_0)v\,(r + \xi\, r_0)$$

  where ξ_i is of order one, this form of the equation of motion says that the net force acting on the particle is the external force reduced by the radiation reaction. This net force at the time $r + \zeta\tau_0$ is equal to mass times acceleration at the time τ. Consequently, it differs from Newton's second law by a time interval between the net force and the inertial term.

Here an essentially new feature emerges, a feature which was not expected and which does not fit into the concepts of classical physics of which this theory is part: the new equation of motion has a non-local behaviour in time, a certain lack of instantaneity which brings with itself a lack of causality over time intervals of the order of τ_{∞}. In particular, energy conservation is no longer satisfied at every instant of time, but is spread out over a time interval of about τ_{∞}. This is a very serious matter and one might be inclined to discard this theory just for that reason alone.

At this point, however, the size of τ_{∞} should be considered. This constant is given in terms of the electron's mass and charge as

$$c_0 = \frac{2}{3}\frac{r_0}{c} = \frac{2}{3}\frac{e}{mc^3} = 6 \times 10^{-24} \text{ sec}.$$

Clearly, such time intervals are entirely outside the domain of competence of classical physics. It follows that the acausality here arises only *outside* the validity limits of the theory.

The situation can now be summarized as follows. By means of a derivation which, because of infinite mass renormalization, etc., must be considered as heuristic, a new equation of motion was obtained. This equation together with the Maxwell-Lorentz field equations define the classical theory of the point-electron. This theory cannot be derived from a Lagrangian or Hamiltonian approach, but is based on these equations as the starting point. Nor does it permit questions about the self-energy or self-stress of the electron.

However, it does describe a point particle and the complete dynamics associated with it. The mass and charge are phenomenological constants which enter the theory.

The theory is classical and has corresponding validity limits. The theory agrees in detail with the requirements of the principle of equivalence. But it shows certain strange non-causal features outside its physical domain of validity which are associated with critical tunes of the order of $\tau_0 = 6 \times 10 - 24$ sec. The theory is mathematically well defined. It agrees with experiment in all cases which have been tested.

Phrased in these terms, a comparison of this theory with the Lorentz-Abraham theory exhibits an interesting instance in the study of scientific theory construction. A typical physical theory (Newtonian mechanics, geometrical optics) has a certain domain of validity outside of which it no longer agrees with experiments.

These limits are usually expressed by inequalities characterizing the accuracy beyond which disagreement will be found. The mathematical structure of the theory does not break down at these limits. In the case of the Lorentz-Abraham theory there are of course also physical validity limits (classical nature of the theory, electron structure not meaningfully described by it).

But in the point electron limit, the theory is mathematically not well defined. Based on the new equation of motion we have a mathematically well-defined theory which exhibits strange physical features outside its validity limits.

This latter fact is mostly irrelevant, but the unambiguous mathematical nature of the theory is essential. Thus, the old theory was completed mathematically and thereby certain new physical features were introduced.

But these lie outside experimental verification, while the experimentally accessible features of the

Lorentz theory have not been changed. This situation is new in the history of theory construction in natural science.

SUCCESSFUL PREDICTIONS AND MATHEMATICAL FAILURE

The discovery of quantum mechanics in 1925 and 1926 called for an extension of the classical theory of the electron to the atomic and sub-atomic domain. At first, however, the electron differed in no way from any other particle which is treated quantum mechanically. As required by experiment, this formalism permits one to explain wave-like as well as particle-like characteristics of the electron.

A most important and unexpected property of electrons was found almost simultaneously with the establishment of this new mechanics; it was discovered that the electron has an intrinsic angular momentum, a spin, and associated with it a magnetic moment. Based on a study of spectra emitted by atoms in a magnetic field, Pauli established a mathematical formalism to account for the apparently dichotomous behaviour of electrons (1925).

Uhlenbeck and Goudsmit thereupon provided the physical picture of a spin angular momentum characterized by ½, i.e.$^{1/(4\pi)}$ times Planck's constant, and two possible orientations.

Essential progress towards a relativistic quantum mechanical description of electrons was made in 1928 when Dirac proposed a new equation which implied a characteristic symmetry property in the sign of the electric charge.

His "hole theory" lead him to the prediction of the existence of positive electrons (first incorrectly identified with protons). Four years later Anderson found the positron in cosmic rays. This symmetry between particles (e.g. "negatons" or negative electrons) and antiparticles (e.g. "positons" or positive electrons) is today one of the most fundamental features of elementary particle theory.

The need for a mathematical formalism that permits not only a quantum mechanical description, but also a mechanism for the creation and annihilation of photons and other particles led in short time to field quantization. The fundamental papers by Heisenberg and Pauli and by Wigner and Jordan provided the basis for the quantum field theory of particles of spin zero and one-half.

While at first charged particles and their electromagnetic interaction were in the centre of interest (quantum electrodynamics), quantum field theory was successfully extended to nuclear interactions by Yukawa (1935-36).

This extension which was actually an analogue of quantum electrodynamics led to the *prediction* of the existence of mesons, i.e. particles which play the same role for nuclear interactions that photons play for electromagnetic interactions. The success of this "meson theory," however, lay mainly in that prediction because, despite great efforts, at best partial and qualitative agreements could be obtained.

The success of quantum field theory was of a very different nature when restricted to quantum electrodynamics. Here excellent experimental agreement resulted from the computations to first approximation of Compton scattering, pair production, bremsstrahlung and other processes. The work of Bethe and Heitler is especially well known.

The attempts to overcome the difficulties of computing effects beyond the first approximation were at first unsuccessful and later interrupted by the second world war. After that, the experiments by Lamb and Retherford ("Lamb shift") signalled the beginnings of new progress (1947). These results posed a great challenge to the theorists: the next approximation was measured for the first time.

The break-through resulted from a combination of two factors. One was the reformulation of quantum field theory in covariant form, which was done independently by Tomonaga and by Schwinger. The other was the appreciation of lessons learned in classical electron theory: Kramers taught mass renormalization to the young American theorists Feynman and Schwinger.

The first results a were the anomalous magnetic moment of the electron (Schwinger, 1948) and an approximate theory for the "Lamb shift" (Bethe, 1947). The latter was then improved by others so as to give the same accuracy as the experiments.

The general formulation, permitting consistent computations to arbitrarily high order was then established by Dyson on the basis of Feynman's diagrammatic methods. Renormalized quantum electrodynamics (within the framework of perturbation theory) was thus established in a few years, although somewhat later contributions proved essential for certain details (Ward and Salam).

The resultant predictive power of the theory is truly remarkable. The energy level difference between the 2s and 2p½ level of Hydrogen was measured to 0.1 Mc/sec accuracy. This means that the corresponding computation relative to the ground state must have an accuracy of about one part in 10. The theory is completely capable of providing the figures to that accuracy. Similar feats relate to the hyper-fine structure and the electron's magnetic moment.

Thus, present quantum electrodynamics is one of the strangest achievements of the human mind. No theory has been confirmed by experiment to higher precision; and no theory has been plagued by greater mathematical difficulties which have withstood repeated attempts at their elimination.

There can be no doubt that the present agreement with experiments is not fortuitous. Nevertheless, the renormalization procedure can only be regarded as a temporary crutch which holds up the present framework. It should be noted that, even if the renormalization constants were not infinite, the theory would still be unsatisfactory, as long as the unphysical concept of "bare particle" plays a dominant role.

If one considers quantum electrodynamics as a phenomenological theory with respect to the mass and charge of the interacting particles, and if one consequently condones the necessity of infinite mass and charge renormalizations, one is tempted to consider quantum electrodynamics as a pretty satisfactory theory.

This can be true only for charged particles of spin one-half, i.e. electrons and mu-mesons.

For the case of charged spinless particles (π- or K-mesons) an essentially new situation arises: the electromagnetic interaction of these mesons diverges, and can be made consistent even within the framework of renormalization theory only by postulating a direct interaction of unknown strength to be determined by experiment. Such an interaction is entirely foreign to Maxwell's equations and can in fact not reasonably be regarded as electromagnetic.

It occurs only in strongly interacting particles and is consequently masked by nuclear interactions. For this reason its determination has so far escaped detection. In fact, it seems not at all possible to separate this direct interaction from a nuclear interaction until a complete theory of nuclear forces is established.

This separation will actually not be possible, because this interaction must be expected to be of nuclear type. Thus, a satisfactory quantum electrodynamics for spinless particles cannot be constructed unless it is concomitant with a theory of strong interactions. Conversely, the theory of nuclear forces can be constructed in a mathematically satisfactory way only when it embraces electrodynamics.

Restricting ourselves now to the theory of electrons and photons, what are the basic difficulties and how do they compare with those of the classical theory? First and foremost, there is the self-energy problem. In the classical theory this problem involved divergent integrals associated with the electromagnetic mass of the point electron, or equivalently, it involved the structure dependence of the whole theory in the case of finite size electrons.

As we have seen, this difficulty can be circumvented by starting the theory not in the Lagrangian or Hamiltonian way, but by means of an equation of motion which contains only the experimental (i.e. renormalized) mass. In quantum electrodynamics the situation is much worse.

The self-energy difficulties are not restricted to the electron, but occur also for the photon. Furthermore, they involve not only divergent integrals associated with the mass, but in fact all integrals (in perturbation theory) which describe dynamical self-interactions, i.e.

the emission and reabsorption of photons, are divergent. These, however contain all observable effects. It is indeed amazing to see an unambiguous procedure by which these meaningless integrals can be separated into another meaningless integral and a convergent one, so that the latter agrees with some of the most accurate experiments ever conducted.

It is possible also in quantum electrodynamics to start with renormalized equations. A formulation of this nature involving the renormalization constants explicitly was given by G. Källen; however, this formulation seems to be inconsistent since it leads to divergences, as was shown by its author.

A formulation which does not involve reference to renormalization was given by Lehmann, Symanzik, and Zimmermann, but it has so far not been carried far enough to serve as a replacement for the present unsatisfactory form of quantum electrodynamics.

Quantum field theory is therefore obviously not in satisfactory mathematical form. It is therefore presently not in a state of completion analogous to the classical theory of the electron.

There are of course many unsolved problems in quantum electrodynamics, but none is comparable to the one just mentioned or should pose a fundamental obstacle.

The problem of obtaining classical electrodynamics, and in particular the integro-differential equation of motion discussed above, as a limit of quantum electrodynamics. That the classical theory must he derivable from the quantum theory is necessitated by the logical structure of physics.

It is furthermore implied by every practicing physicist who uses the classical theory for the construction of accelerators (orbit calculations) and the quantum theory for the prediction of the experiments he plans to conduct with these machines. This problem is a very essential logical link in the theory of the electron.

The basic remaining problem, however, is that of a reformulation, so as to clad the completely satisfactory experimental predictions into a clean mathematical structure. It is my belief that considerable progress in that direction can be made by trying to bring the formalism closer to the physical situation.

This idea has in fact been fruitful recently when it was possible to eliminate certain mathematical complexities by means of it. The case in point is the following. The usual formulation of quantum electrodynamics is based on the potential four-vector as the fundamental quantity which describes the electromagnetic field.

This has partly historical reasons: In classical electrodynamics it was found that the Maxwell equations for the field strengths can be greatly simplified if one introduces the potentials (scalar and vector potential) from which the field strengths can be derived. It was realized that these potentials have no direct physical significance and only field strengths can be measured; but it was a small price to pay for the great mathematical simplification.

When these potentials are introduced in classical electrodynamics, they bring with themselves superfluous degrees of freedom. However, these are easy to eliminate by means of the Lorentz condition and the requirement of gaugeinvariance.

In quantum electrodynamics this same introduction of potentials leads to various complications. Consistency with the quantization conditions (commutation relations) requires modifications of the Lorentz condition. Further difficulties arise and further mathematical complications (indefinite metric) are required.

Thus, by the time a consistent formulation is achieved, the simplification for which the potentials were originally intended has been completely off-set by a rather formidable mathematical apparatus. At the same time, all these complications refer to the unobservable potentials and are therefore of no physical relevance. Recently, it has been possible to reformulate quantum electrodynamics completely in terms of the field strengths of the electromagnetic radiation.

The static and quasi-static interactions between moving charges are retarded and are of a classical nature: they are not associated with photons. Only the radiation fields are quantized. They lead to transverse photons only and no other photons occur in the theory. In this form the theory is free of questions of gaugeinvariance and no supplementary condition need be satisfied. These are tremendous simplifications. At the same time, many unphysical features are eliminated.

GUESSING FUTURE DEVELOPMENTS

At the present time no one can tell what a future mathematically satisfactory quantum electrodynamics will be like. But there are many guesses which have their origin in the relative importance various physicists ascribe to those principles and prejudices on which the present theory is based. Since a solution of this problem may not come within the immediate future, it seems safe to add my own ideas on this matter. The theory of the electron does not seem to be "coupled" to the strong interactions in contradistinction to the theory of charged boson, especially of the π-and K-mesons.

Furthermore, since the reduction of classical to quantum electrodynamics must surely exist and the latter is presently in relatively satisfactory shape, various characteristic features can be gleaned from it. First, there is a strong hint that Lagrangian and Hamiltonian formulations are not essential and in fact not desirable: the classical Lorentz-Dirac equation cannot be derived from a Lagrangian.

Secondly, the asymptotic properties of the system (both in time and in space) play a much more fundamental role than has hitherto been realized.

In general relativity a similar emphasis pervades recent work. The asymptotic formulation of field theory, begun by Lehmann, Symanzik, and Zimmermann seems therefore very much in the right direction. Thirdly, classical electrodynamics teaches us that the theory in the small will most likely be of a non-local nature.

This is confirmed by the formulation of quantum electrodynamics in terms of field strengths: Some of the important quantities of the theory are given by integrals over all space and are not point functions. Further confirmation of this conjecture is obtained from a renormalization of the theory in this formulation: the renormalized commutation relations are non-local; in particular, they no longer vanish outside the light cone (micro-causality).

Only asymptotically do these commuters vanish (asymptotic or macro-causality). This feature is again closer to the experimental situation than the local field theory: quantum mechanical measurements are necessarily restricted to averages over finite (though small) space-time regions and strict micro-causality is not observable in principle.

Fourthly, associated with the non-local nature, as it emerges here, is a lack of local Lorentz invariance. Various quantities depend on space-like planes and are no longer scalars or tensors as in the local theory. Only in the distant past and future (incident and outgoing particles) are the transformation properties restored. Thus, we have asymptotic Lorentz invariance. It is quite possible that other invariance properties of the local theory will also become valid only asymptotically.

These are the features which seem to emerge and point toward the possible characteristics of the future quantum electrodynamics and perhaps of quantum field theory in general. They can be generally described as a fuzziness and indeterminacy for small space-times, with a consequent blurring of detail in the microscopic description, while leaving all known features unchanged asymptotically. The macroscopic theory is unchanged.

It is possible and in fact likely that some basic new principles and ideas will have to be added to what will be left of standard quantum field theory in order to provide for a complete and consistent description of electromagnetic interaction, and, in particular, of the electron. The future will tell which of these various conjectures are valid. If, nineteen years from now, the fiftieth Joseph Henry Lecturer should again discuss the theory of the electron, the chances are good that he will know the answer to many of these questions.

ELECTRON THEORY OF METALS

Basic Definitions Relating to the Motion of Free Electrons in a Metal:

(*i*) *Random Velocity (Vran):* In the absence of external Electric field E, the free electrons in a conductor (metal) move randomly in all possible directions undergoing scattering at the ionic sites. The velocity of these randomly moving electrons in the absence of applied electric field is called Random Velocity 'Vran'. The current due to the random motions of these free electrons is Zero.

(*ii*) *Drift Velocity (Vd): In* the presence of external Electric field E, the average velocity acquired by free electrons between two successive collisions and has acquired a steady state in a direction opposite to the direction of applied field is called Drift Velocity 'Vd'.

(*iii*) *Mean free path (l):* In the presence of external Electric field E, the average distance travelled by a free electron between two successive collisions is called Mean free path 'l'.

(iv) *Relaxation time (t):* In the presence of external Electric field E, the time taken by an electron to reach its equilibrium state from its disturbed state is called Relaxation time 't'.

(Or)

In the absence of external electric field, after an electron reaches the stead state velocity, the time taken for the drift velocity to decay to 1/e times of its initial value is called the relaxation time 't'.

(*v*) *Mobility* (μ): In the presence of external Electric field, the magnitude of drift velocity |Vd | per unit Electric field E is defined as Mobility of electrons 'μ'.

(*vi*) *Current Density* (*j*): It is defined as the ratio of the current 'i' to the cross-sectional area 'A' of the given conductor

(Or)

It is the amount of charge '-*n*e' crossing a unit area 'A' per unit time 't'.

Classical Free Electron Theory or Drude-Lorentz theory of Metals

This theory is proposed by Lorentz and Drude in 1900. The conducting properties of a solid are not a function of the total number of electrons in the solid. This is because; only the outermost electrons of the atoms take part in conduction.

Postulates of Classical Free Electron theory

1. The outermost electrons of an atom are not bound to that atom, but are free to move through the whole solid like the molecules of a perfect gas in a container. These electrons are also called as the free electron gas or free electron cloud.
2. Collisions made by these electrons either with positive ions or with other electrons are of elastic in nature. That is, no loss of energy during collisions.
3. The movements of these electrons obey the laws of Classical Kinetic Theory of Gases.
4. The velocities of these electrons obey the Classical Maxwell-Boltzmann Distribution of velocities.
5. These electrons move in Uniform potential field due to the positive ions fixed in the lattice.
6. In presence of external electric field, these electrons move in a direction opposite to the direction of the field.

Merits of Classical Free Electron Theory

1. It successfully verifies Ohm's Law.
2. It explains the electrical and thermal conductivities of metals.
3. It derives Wiedemann-Franz Law (the ratio of Thermal conductivity to the Electrical conductivity is a constant i.e. K/s = Constant).

Expression for Electrical Conductivity in Terms of Number of free Electrons 'n' and their Mobility 'μ'

(Verification of Ohm's Law): In the absence of applied electric field, the net current due to the movement of electrons is Zero since they move randomly in all directions. The current in a conductor is due to the motion of free electrons under the influence of the external electric field 'E'. emerits of Classical free electron theory:

In spite of the above successful explanation of the electrical conductivity, Classical theory has the following drawbacks.

1. It failed to explain the phenomena such as Photoelectric effect, Compton effect and the Black body radiation.
2. Electrical conductivity of semiconductors couldn't be explained using this model.
3. Ferromagnetism couldn't be explained by this theory.
4. According to this theory, K/sT=constant (Wiedemann-Franz law) but it is not a constant at low temperatures.

QUANTUM FREE ELECTRON THEORY OR SOMMERFELD'S THEORY OF METALS

Sommerfeld in 1928 investigated the behaviour of a free electron gas taking into account the Quantum theory and the Pauli's exclusion principle.

Postulates of Quantum Free Electron Theory:

1. Electrons are considered to be subjected to a constant potential 'V'.
2. The collisions of these electrons with the other electrons and with the positive ions are negligible i.e. they are free from frictional forces.
3. Electrons can only take discrete energy values but not continuous and follow Pauli's exclusion principle.
4. Electrons in a metal follow the Fermi-Dirac statistics and hence they are treated as Fermi gas.

Expression for the electrical conductivity by applying Quantum laws: When an external electric field 'E' is applied, the force 'F' exerted on the electron is given by

$$F = -eE$$

According to Newton, force is also rate of change of momentum given by

$$F = \frac{dp}{dt}$$

Hence the equation governing the motion of an electron is given by

$$-eE = \frac{dp}{dt}$$

But according to de broglie, every material particle is associated with a wave whose wave length is given by

$$\lambda = \frac{h}{p} \text{ or } p = \frac{h}{\lambda}$$

$$p = \frac{h}{2\pi} \times \frac{2\pi}{\lambda}$$

$$p = \text{hK}$$

Where 'K' is a wave vector of magnitude $\frac{2\pi}{\lambda}$

Substituting equation, we have

$$-eE = \frac{d\hbar K}{dt} = \hbar\frac{dK}{dt} \Rightarrow dK = \frac{-eE}{\hbar}dt$$

The above equation represents that the origin of a K space moves through a distance dK on time dt on application of external field.

In steady state $dK = \Delta\text{K}$ and $dt = \tau$ the average collision time.

Hence
$$\Delta K = \frac{-eE}{\hbar}\tau$$

We know that $p = mv = hK$, then incremental velocity Δv is given by

$$\Delta v = \frac{\hbar}{m}\Delta K$$

$$= -\left(\frac{\hbar}{m}\right)\left(\frac{eE\tau}{\hbar}\right)$$

$$\Delta v = -\frac{eE\tau}{\hbar}$$

If represents 'n' the electron density then current density J in terms of electron density is given by

$$J = n(-e)\,\Delta v$$

$$= \frac{ne^2E\tau}{m}$$

$$= \left(\frac{ne^2\tau}{m}\right)E$$

$$J = \sigma E \text{ (Ohm's Law)}$$

Fermi-dirac Distribution Function F(E): Fermi-Dirac Distribution function gives the distribution of electrons in real material among the different possible energy states. We know that an assembly of electrons may be considered as Fermi gas which obey Fermi-Dirac statistics.

The probability F(E) of an electron occupying an energy level E is given by

$$F(E) = \frac{1}{1 + \exp\left(\frac{E - E_F}{kT}\right)}$$

Case (i): At absolute temperature i.e. T = 0K and for energies below the Fermi energy level i.e. E < EF, the above equation results as shown below

$$F(E) = \frac{1}{1 + \exp(-\infty)} = \frac{1}{1+0} = 1$$

Equation says that, the probability of finding an electron in the energy level below the Fermi level is maximum.

Case(ii): At absolute temperature i.e. T = 0K and for energies above the Fermi energy level i.e. E > EF, equation (1) results as below

$$F(E) = \frac{1}{1 + \exp(\infty)} = \frac{1}{\infty} = 0$$

Equation says that the probability of finding an electron in the energy level above the Fermi level is Zero. From this we conclude that *"the Fermi energy level is the maximum possible energy level that an electron can occupy at absolute temperature"*. In other words, at absolute temperature, all energy levels below the Fermi energy level are completely filled whereas all energy levels above the Fermi level are completely vacant.

General case: At temperatures other than absolute zero i.e. T > 0K, and in the energy level equal to Fermi level i.e. E = EF, equation results as below

$$F(E) = \frac{1}{1 + \exp(0)} = \frac{1}{2} = 0.5$$

From equation Fermi energy level may be defined as *" it is the energy level for which the probability of finding an electron is half at temperatures greater than absolute zero". Graphical representation of Fermi distribution function:*

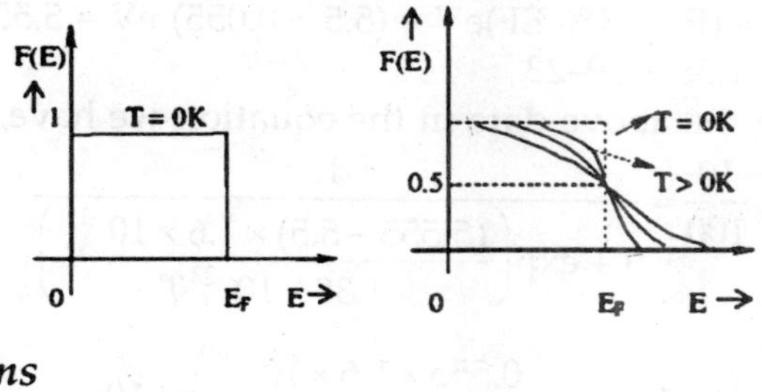

Solved Problems

1. Find the temperature for which there exists 1% probability of an

electron occupying an energy level at 0.5eV above the Fermi level.

Soln: We know that the probability function is given by

$$F(E) = \frac{1}{1 + \exp\left(\frac{E - E_F}{kT}\right)}$$

Given F(E) = 1%

E = (EF + 0.5) eV = (EF + 0.5) × 1.6 × 10–19 J.

k 1.38 × 10–23

Substituting the above data in the equation we have,

$$\frac{1}{100} = \frac{1}{1 + \exp\left(\frac{(E_r + 0.5 - E_r) \times 1.6 \times 10^{-19}}{1.38 \times 10^{-23} T}\right)}$$

$$1 + \exp\left(\frac{(E_r + 0.5 - E_r) \times 1.6 \times 10^{-19}}{1.38 \times 10^{-23} T}\right) = 100$$

$$\frac{0.5 \times 1.6 \times 10^{-19}}{1.38 \times 10^{-23} T} = \ln 99$$

$$T = \frac{0.5797 \times 10^4}{\ln 99} = 1261.6K$$

2. At what temperature we can expect a 10% probability that electrons in silver have an energy which is 1% above the Fermi energy? The Fermi energy of Silver is 5.5eV.

Solution: We know that the probability function is given by

$$F(E) = \frac{1}{1 + \exp\left(\frac{E - E_F}{kT}\right)}$$

Given F(E) = 10% = 0.1

EF = 5.5eV

EF = 5.5eV

E = (EF + 1% EF)eV = (5.5 + 0.055) eV = 5.555 eV

K 1.38 × 10–23

Substituting the above data in the equation we have,

$$\frac{10}{100} = \frac{1}{1 + \exp\left(\frac{(5.555 - 5.5) \times 1.6 \times 10^{-19}}{1.38 \times 10^{-23} T}\right)}$$

$$1 + \exp\left(\frac{0.555 \times 1.6 \times 10^{-19}}{1.38 \times 10^{-23} T}\right) = 10$$

$$\frac{637.68}{T} = \ln 9 \Rightarrow T = \frac{637.68}{\ln 9} \Rightarrow T = 290.2K$$

Density of Energy States And Carrier Concentration: The number of electrons per unit volume in a given range (between E and E + dE) of energy or Carrier concentration is given by summing the product of the density of energy states Z(E) and the probability of occupancy F(E), i.e.

$$n_C = \int_{\text{energy band}} Z(E)F(E)dE$$

$$Z(E)dE = \frac{\pi}{2}n^2 dn$$

We know that En = $\frac{n^2 h^2}{8ma^2}$

Rearranging the above equation, we have

$$n^2 = \frac{8ma^2}{h^2}E_n$$

$$n = \left(\frac{8ma^2}{h^2}\right)^{1/2} E^{1/2}$$

Differentiating the quation (4),

$$2n\, d_n = \frac{8ma^2}{h^2}dE$$

$$d_n = \frac{1}{2n}\frac{8ma^2}{h^2}dE$$

Substituting the value of 'n' from equation in the above equation, we have

$$dn = \frac{1}{2}\left(\frac{8ma^2}{h^2}\right)\left(\frac{h^2}{8ma^2}\right)^{1/2}\frac{1}{E^{1/2}}dE$$

$$= \frac{1}{2}\left(\left(\frac{8ma^2}{h^2}\right)\frac{h^2}{8ma^2}\right)^{1/2}\frac{1}{E^{1/2}}dE$$

$$dn = \frac{1}{2}\left(\frac{8ma^2}{h^2}\right)^{1/2}\frac{1}{E^{1/2}}dE$$

Now substituting the values of 'n' & 'dn' from equations (5) & (8) in equation we have

$$Z(E)\ dE = \frac{\pi}{2}\left(\frac{8ma^2}{h^2}E\right)\left(\frac{1}{2}\left(\frac{8ma^2}{h^2}\right)^{1/2}\frac{1}{E^{1/2}}\right)dE$$

$$= \frac{\pi}{2}\left(\frac{8ma^2}{h^2}\right) E^{1/2} dE$$

$$Z(E)\ dE = \frac{\pi}{2}\left(\frac{8m}{h^2}\right)^{3/2} a^3 E^{1/2} dE$$

But according to Pauli's Exclusion principle two electrons of opposite spin can occupy each state and hence the number of energy states available for the electron occupancy is given by

$$Z(E)\ dE) = 2 \times \frac{\pi}{2}\left(\frac{8m}{h^2}\right)^{3/2} a^3 E^{1/2} dE$$

equation represents the Density of energy states in a given volume a^3. Therefore the density of energy states per unit volume is given by

i.e. Density of states $Z(E)\ dE = \frac{4\pi}{h^3}(2m)^{3/2} E^{1/2} dE$

We know that the probability of occupancy of an electron in the given energy state is given by

$$F(E) = \frac{1}{1 + \exp\left(\frac{E - E_r}{kT}\right)}$$

Substituting the values of Z(E) & F(E) from equations, we have

$$nC = \int_{Energy\ band} \frac{4\pi}{h^3}(2m)^{3/2} E^{1/2} = \frac{1}{1 + \exp\left(\frac{E - E_F}{kT}\right)} dE$$

$$nC = \frac{4\pi}{h^3}(2m)^{3/2} \int_{Energy\ band} E^{1/2} = \frac{1}{1 + \exp\left(\frac{E - E_F}{kT}\right)} dE$$

equation is the expression for *Density of charge carriers or carrier concentration* in a given metal at any temperature.

ELECTRON SCATTERING AND RESISTANCE OR SOURCES OF ELECTRICAL RESISTANCE

The main factors affecting the electrical resistance in solids are temperature and defects or impurities present in the solid.

According to Matthiessen's rule, the total resistivity of the material is given by

$$\rho_{oeat} = \rho_{pare} + \rho_{imparicy}$$

where ρ_{pare} is temperature dependent resistivity lattice while $\rho_{imparicy}$ is due to the scattering of electro material.

Temperature dependence of resistivity, ρ_{pure}:

We know that as the temperature 'T' of the material increases the amplitude 'A' of the lattice vibrations increases, according to $A^2 \alpha kT$. Hence these lattice vibrations cause the scattering effect on the electrons.

This scattering is the cause of electrical resistance in metals. Also, the mean free path 'λ' of the electrons decreases with the increase in amplitude 'A' of the lattice vibrations according to the relation $\lambda \alpha \frac{1}{A^2}$.

Hence, at room temperature $\rho \alpha T \alpha \frac{1}{\lambda}$ but at low temperature r a Ta where n ~ 3 to 5 which implies that as $T \rightarrow 0$, $\rho \rightarrow 0$ where the material behaves as a superconductor.

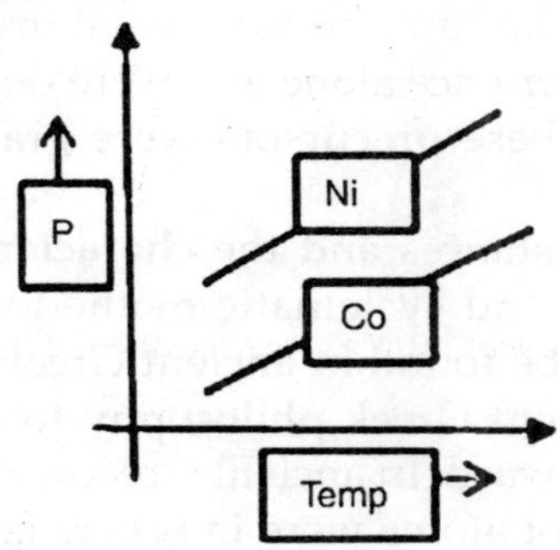

FORMATION OF THE ATOMIC THEORY

Modern chemistry is based on the atomic theory. To understand the atomic theory, you must first learn the fundamental laws of chemistry. Such fundamental laws include the law of conservation of mass, the law of definite proportions, and the law of multiple proportions. These are the basis of the atomic theory and at the same time represent conclusions drawn from the atomic theory. However, the atomic theory by itself was incomplete. Chemistry could be a consistent system for the first time only when the atomic theory was combined with the concept of molecules. In the past, the existence of the atom was a mere hypothesis.

At the beginning of the 20th century was the atomic theory finally proved. It also became clear that the atom itself consisted of several smaller particles. Current atomic theory gradually developed along this line and constituted the framework of the material world.

The Birth of Chemistry

Modern chemistry was initiated by the French chemist Antoine Laurent Lavoisier (1743-1794). He established the law of conservation of mass for chemical reactions, and revealed the role of oxygen in combustion. Based on these principles, chemistry progressed in the right direction. Actually oxygen was discovered independently by two chemists, the English chemist Joseph Priestley (1733-1804) and the Swedish chemist Carl Wilhelm Scheele (1742-1786), at the end of the 18th century. Thus, only about two hundred years have passed since modern chemistry was born. In this regard, chemistry is a relatively young science as compared with physics and mathematics, both of which have a history of several thousand years.

However, alchemy, metallurgy and pharmacy in ancient times might be regarded as the roots of chemistry. Many discoveries by those who engaged in these fields contributed to modern chemistry although alchemy was founded on a false theory.

Moreover, prior to the 18th century, metallurgy and pharmacy were essentially based on experience alone and were devoid of theory. Thus, it seems impossible that these precursors were gradually developed into modern chemistry.

Given these circumstances and the character of modern chemistry with its well-organized and systematic methodology, the true roots of modern chemistry may be found in ancient Greek philosophy.

The road from ancient Greek philosophy to modern atomic theory was not at all straightforward. In ancient Greece, atomic theory as well as denials of the existence of atoms were in severe conflict.

In fact, atomic theory long remained unorthodox in the world of chemistry and science. Men of learning were not interested in atomic theory until the 18th century. At the beginning of the 19th century, the English chemist John Dalton (1766-1844) revived the atomic theory of ancient Greece.

Even after this revival, not all scientists necessarily approved of the atomic theory. Not until the beginning of the 20th century the atomic theory was finally proved as a fact, not a mere hypothesis. It was accomplished by skilful experiments of the French physicist Jean Baptiste Perrin (1870-1942).

Thus, it took a long time for the atomic theory to be established as the basis of modern chemistry. As noted earlier, chemistry is a relatively young science. Consequently, much remains to be done before chemistry can claim to fully understand matter, and through its understanding of matter, nature itself. So, it is very important upon beginning to study chemistry that we review briefly how chemistry has developed since its birth.

(a) Ancient atomic theory

As stated above, the roots of modern chemistry lie in the atomic theory developed by ancient Greek philosophers. The philosophy of ancient

Greek atomism is attributed to Democritos (*ca.* 460 BC-*ca.* 370 BC). However, no writings of Democritos are extant. Therefore, our resource must be the long poem *"De rerum natura"* written by the Roman poet Lucretius (*ca.* 96 BC-*ca.* 55 BC).

The atom described by Lucretius has some resemblance to the modern molecule. Wine and olive oil, for instance, had their own atoms. The atom was not an abstract entity. Rather, the atom had a definite shape with a function that appropriately corresponded to its shape. "The atom of wine is round and smoothly passes through your throat while the atom of wormwood is jagged and scratches your throat." Modern structural theory of molecules says that there is a very close relation between the structure of a molecule and its functions.

Though the philosophy as articulated by Lucretius was not supported by evidence gained by experiments, it was a precursor of modern chemistry.

(a) (b)

Fig. The World of Democritos.

We unfortunately cannot guess the image of atoms as conceived in Democritos' thought. A German chemist has suggested an image of atoms as they might have been conceived by Democritos.

(a) An atom of sweet substance.

(b) An atom of bitter substance. (reproduced from: F. Berr, W. Pricha, *Atommodelle,* Deutsches Museum, 1987.

In the long period ranging from ancient times to the end of the medieval age, the atomic theory remained heretical because the theory of four elements (water, earth, air and fire) by the ancient Greek philosopher Aristotole (384 BC-322 BC) reigned. When the authority of Aristotle declined at the beginning of the modern age, many philosophers and scientists began to develop theories influenced by ancient Greek atomic theory. The images of matter held by the French philosopher Rene Descartes (1596-1650), the German philosopher Gottfried Wilhelm Freiherr von Leibniz (1646-1716), and the English scientist Sir Issac Newton (1642-1727) were more or less influenced by that atomic theory.

(b) Dalton's Atomic Theory

At the beginning of the 19th century, atomic theory as a philosophy of matter was already well developed by Dalton who developed his atomic theory based on the role of atoms during chemical reactions. His atomic theory may be summarized as below.

Dalton's atomic theory:

(i) The ultimate particle which composes the elements is the atom. All atoms of a particular element are identical.

(ii) The masses of atoms of the same type are equal but that of the other types are different.

(iii) The whole atom is involved in a chemical reaction. The whole atom forms a part of a compound. The type and number of atoms in a given compound are fixed.

The theoretical foundation of Dalton's theory was mainly based on the law of conservation of mass and the law of definite proportiona), both of which had already been established, and the law of multiple proportion) which was developed by Dalton himself.

The atom of Democritos may be said to be as a kind of miniature of matter. Hence the number of the type of atoms is equal to that of the type of matter.

On the other hand, Dalton's atom is a constituent of matter, and many compounds are formed by the combination of a limited number of atoms. Hence, there should be a limitation in the number of types of atoms. Dalton's atomic theory requires the process in which two or more atoms combine to form matter. This is the reason why Dalton's atom is called the "chemical atom".

(c) *Proof that atoms exist:* When Dalton initially proposed his atomic theory, it attracted some attention. However, it failed to gain unanimous support. Some supporters of Dalton made considerable efforts to persuade those who were against the theory, but some opposition remained. The chemistry of his day was not sufficient to prove the existence of atoms by experiments.

So the atomic theory remained a hypothesis. Moreover, science after the 18th century developed many experiments that made scientists were very skeptical of the hypothesis. For example, such noted chemists as Sir Humphry Davy (1778-1829) and Michael Faraday (1791-1867), both English, were skeptical towards the atomic theory.

While the atomic theory remained a hypothesis, great advances were made in the field of science. One was the rapid rise of thermodynamics in the 19th century.

Structural chemistry of the day as represented by the atomic theory was purely academic with little possibility of practical application. But thermodynamics derived from such a practical issue as the efficiency of steam engines loomed more important. There was a very severe controversy between atomists and those who supported thermodynamics.

The debate between the Austrian physicist Ludwig Boltzmann (1844-1906) and the German chemist Friedrich Wilhelm Ostwald (1853-1932) and with the Austrian physicist Ernst Mach (1838-1916) is noteworthy. It had a grave consequence, Boltzmann's suicide.

At the beginning of the 20th century, there was a great change in the popular interest in science. A series of important discoveries, including the discovery of radioactivity, aroused interest in the nature of atoms, and more generally, in structural science.

That atoms exist was experimentally confirmed by the sedimentation equilibrium experiments carried out by Perrin The English botanist Robert Brown (1773-1858) discovered an irregular motion of colloidal particles, and this movement was called Brownian motion after him. The Swiss physicist Albert Einstein (1879-1955) developed a theory of this motion based on the atomic theory.

According to this theory, the Brownian motion could be expressed by an equation including Avogadro's number

$$D = \frac{RT}{N}\frac{1}{6\pi\alpha\eta}$$

where D is the movement of the particles, R the gas constant, T the temperature, N Avogadro's number, the radius of the particles andçthe viscosity of the solution.

The essence of Perrin's idea is as follows. Colloidal particles move randomly by Brownian motion and simultaneously sink downward by gravitation. Sedimentation equilibrium results from the balance of these two motions, random motion and sedimentation by gravity.

Perrin carefully observed the distribution of colloidal particles, and with the aid of Eq. 1.1 and this data, he obtained Avogadro's number. Surprisingly the value agreed well with Avogadro's number estimated by completely different methods. This agreement in turn proved the correctness of the atomic theory which was the basis of the theory of Brownian motion.

Needless to say, Perrin could not observe atoms directly. What scientists of those days, including Perrin, could do is to demonstrate that Avogadro's number obtained from various different methods based on the atomic theory is always identical. In other word, they proved the atomic theory rather indirectly by logical consistency.

In the frame of modern chemistry, this kind of methodology is still important. Even today it is impossible to observe directly such small particles as atoms with naked eyes or an optical microscope. To be observed directly by visible light, the size of the particles should be larger than the wavelength of visible light.

The wavelength of visible light is in the range of 4.0×10^{-7}- 7.0×10^{-7} m, which is *ca.* 1000 times larger than the size of atoms.

Hence it was beyond the range of optical instruments to observe atoms. With the aid of new devices such as the electron microscope (EM) or scanning tunneling microscope (STM), that impossibility was overcome. Though the principle of observing atoms by these devices is, however, different from what is involved in our observing the moon or flowers, we may be allowed to say that we can now observe the image of atoms almost directly.

THE COMPONENTS OF MATTER

(*a*) *Atoms:* In the world of chemistry based on the atomic theory, the ultimate smallest unit that constitutes matter is the atom. Matter is defined as a sum of atoms. The atom is the smallest component of an element that will not suffer from any change during a chemical reaction. All atoms are composed of common components, a nucleus and electrons. The diameter of a nucleus is *ca.* 1/10,000 that of an atom. More than 99 % of the mass of an atom is concentrated in the nucleus.

The nucleus is composed of proton(s) and neutron(s), and their numbers determine the property of the element. The mass of a proton is *ca.* 1.67×10^{-27} kg and bears a unit positive charge, 1.60×10^{-19} C (Coulomb). This amount is the smallest unit of electric charge and called the elementary electric charge. A nucleus has a positive electric charge whose amount depends on the number of protons it contains. The mass of a neutron is much the same as that of a proton, but a neutron does not possess any electric charge. An electron is a particle with unit negative charge, and a given atom contains the same number of electrons as that of protons in its nucleus.

Hence an atom is electrically neutral. The properties of these particles that constitute atoms are summarized in Table 1.1. The number of protons in the nucleus is called the atomic number and the sum of the number of protons and the number of neutrons is called the mass number. Since a proton and a neutron have much the same mass, and the mass of the electron is negligible as compared with those of a proton and a neutron, the mass of an atom is almost proportional to its mass number. When atomic number and mass number of a given atom are to be indicated, the atomic number is added to the lower left of the symbol of atoms as a subscript, and the mass number to the upper left of it as a superscript.

Table. Properties of Particles that Constitute Atoms

	mass(kg)	*relative mass*	*electric charge (C)*
proton	1.672623 × 10–27	1836	1.602189 × 10–19
neutron	1.674929×10–27	1839	0
electron	9.109390 × 10–31	1	–1.602189×10–19

The number of protons and electrons possessed by an element determines the chemical properties of that element. The number of neutrons may vary. A given element will always have the same atomic number but may have different numbers of neutrons, and hence a different mass number. These variants are called isotopes. As an example, the isotopes of hydrogen are listed in the table

Table: Isotopes of hydrogen

symbol and name	*mass(kg)*	*relative mass*
1H hydrogen	1	0
2H deuterium, D	1	1
3H tritium, T	1	2

(*b*) *Molecules:* The smallest component of independently existing neutral matter is called molecule.

Monatomic molecules are composed of one atom (*e.g.*, Ne). Polyatomic molecules are composed of more atoms (*e.g.*, CO2). The kind of bond between atoms in polyatomic molecules is called a covalent bond.

One reason why it took some time before atomic theory was fully accepted is as follows. In his theory Dalton admitted the existence of molecules (in modern terminology) which are formed by combinations of atoms of different types, but he did not accept the idea of diatomic molecules for elements such as oxygen, hydrogen or nitrogen which were being extensively investigated at that time. Dalton believed in the so-called "principle of the simplest" a) and based on this, he automatically assumed that elements such as hydrogen and oxygen are monatomic.

The French chemist Joseph Louis Gay-Lussac (1778-1850) proposed a law for gaseous reactions which said that in gaseous reactions, the volume ratio was a whole number. Dalton's atomic theory could not give rationale for the law. In 1811, the Italian chemist Amedeo Avogadro (1776-1856) proposed that gaseous elements such as hydrogen and oxygen were not monatomic but diatomic. Furthermore, he also proposed that at constant temperature and pressure, all gases contained the same number of particles in a given volume. This hypothesis was initially termed Avogadro's hypothesis, but later was called Avogadro's law.

Avogadro's law could provide the basis for determining the relative masses of atoms, that is, the atomic weights (traditionally "weight" has been used instead of "mass"). Its significance was only tardily granted. The Italian chemist Stanislao Cannizzaro (1826-1910) realized the significance of Avogadro's hypothesis and strongly advocated its validity at the first International Chemical Congress held in Karlsruhe, Germany, in 1860, which was called to discuss international agreement for standard atomic weights. Thereafter, the validity of Avogadro's hypothesis was gradually accepted.

(*c*) *Ions:* An atom or a group of atoms bearing an electric charge is called an ion. A cation is an ion having a positive charge, an anion one having a negative charge. An electric attraction exists between a cation and an anion. In the crystal of sodium chloride (NaCl), sodium ion (Na^+) and chloride ion (Cl^-) are bonded with that electric attraction. This kind of bond is called an ionic bond.

STOICHIOMETRY

(*a*) *The early stage of stoichiometry:* At the beginning of chemistry, the quantitative aspect of chemical change, *i.e.*, the stoichiometry of a chemical reaction, was not given much attention. Even if some consideration were given, primitive experimental devices and techniques did not yield correct results.

One example involves the phlogiston theory. Phlogistonists attempted to explain the phenomenon of combustion in terms of "combustible earth". According to them, combustion was the release of this combustible earth (from the combustible substance). It was later called "phlogiston". Based on this theory, they defined combustion as the discharge of phlogiston from combustible matter. The change of mass of wood when it was burnt agreed nicely with this theory.

However, the change of mass of metals when calcinated disagreed with the theory. In spite of that phlogistonists accepted that the nature of the two processes were essentially identical. The increase of mass of calcinated metals was a fact. The phlogistonists tried to explain this anomaly by assigning to phlogiston a negative mass. The Flanders philosopher Jan Baptista van Helmont (1579-1644) carried out a famous "willow" experiment. He grew a seedling of willow after measuring the combined weight of flowerpot and soil. Since there was no change in the weight of the flowerpot and the soil while the seedling grew up, he considered that the gain in weight was due only to the water he gave to the seedling.

He concluded that "the root of all matter is water". Judged from today's viewpoint, his hypothesis and experiment were far from perfect, but this story is a good example of an emerging attitude for the quantitative aspects of chemistry. Helmont recognized the importance of stoichometry, and in that regard he was well ahead of his times.

At the end of the 18th century, the German chemist Jeremias Benjamin Richter (1762-1807) established the concept of equivalent (in modern terminology chemical equivalent) by careful examination of acid/base reactions, that is, the quantitative relation between the acid and the base in neutralization reactions. His equivalent, or now chemical equivalent, indicates a certain amount of matter in the reaction. An equivalent in neutralization corresponds to the relation between a given amount of acid and the amount of base to neutralize it. Knowledge of the correct equivalent was essential in order to produce good soap or gunpowder. Thus, such knowledge was of great practical importance. At the same time Lavoisier established the law of conservation of mass, and gave a basis for the concept of equivalent by his ingenious and accurate experiments. Thus, stoichiometry which handles the quantitative aspects of chemical reactions became the fundamental methodology of chemistry.

All fundamental laws of chemistry, from the laws of conservation of mass, of definite composition and of multiple proportions to the law for gaseous reactions, are all based on stoichiometry. These fundamental laws were the basis of atomic theory, and were consistently explained in terms of atomic theory. It is interesting to note, however, that the concept of equivalent was used before the atomic theory was introduced.

(*b*) *Relative masses of atoms and atomic weights:* Dalton recognized that it was necessary to determine the weight of each atom since it varied for each type of atom. Atoms were so small that it was impossible to determine the weight of one atom. So, he focused on the relative values of the weight and made a table of atomic weights for the first time in human history. In this table, the weight of the lightest element, hydrogen, was set to unity as the standard (H = 1). Atomic weight is a relative value, *i.e.,* a ratio without any dimension. Though some of his atomic weights were differed from modern values, most of his values were in the allowable range. This in turn shows that his ideas and experiments were correct.

(*c*) *Chemical formula weight and molecular weight:* Each compound is defined by a chemical formula indicating the type and the number of atoms constituting that compound. Formula weight (or chemical formula weight) is defined as the sum of atomic weight based on the type and the number of atoms defined by the chemical formula. The chemical formula of a molecule is called molecular formula, and its chemical formula weight is termed molecular weight. (a) For instance, the molecular formula of carbon dioxide is CO_2, and the molecular weight is 12 + (2 × 16) = 44. As with atomic weights, both formula weight and molecular weight are not necessarily integers. For instance, the molecular weight of hydrogen chloride HCl is 36.5. Even if the type and the number of atoms constituting the molecule are identical, the two molecules may have different molecular weights if different isotopes are involved.

The difference in molecular weight of H_2O and D_2O is substantial, and the differences in physical and chemical properties between these two compounds are not negligible. H_2O is more readily electrolyzed than D_2O.

Thus, the water remaining after electrolysis tends to contain more D_2O than natural water.

(*d*) *Quantity of matter and mole:* The most suitable method of quantitatively expressing the amount of matter is the number of particles such as atoms or molecules which constitute the matter in question. However, to count such small and invisible particles as atoms or molecules would be very difficult. Instead of counting the number of particles directly, we can use the mass of a definite number of particles.

Then, how was such a definite number chosen? To make a long story short, the number of particles contained in a gas of 22.4 dm^3 at standard pressure and temperature was chosen as this standard number. This number is called Avogadro's number for historical reasons. The name Loschmidt number was also proposed after the name of the Austrian chemist Joseph

Loschmidt (1821-1895) who first determined this number by experiment (1865).

Definition of "mol":

(*i*) The amount of matter that contains the same number of particles as that contained in 12 g of 12C.

(*ii*) One mol of matter contains the same number of particles as Avogadro's constant.

(*iii*) The amount of matter constituted in 6.02×10^{23} particles is one mol.

Since 1962, according to SI (Systeme Internationale) it was decided that, in the chemical world, mole (abbreviation mol) is to be used as the unit of the amount of matter. Avogadro's number was defined as the number of carbon atoms contained in 12 g of ^{12}6C and renamed Avogadro's constant. There are several definitions of "mol".

(*e*) *Atomic mass unit (amu)*: As the standard of the mass of atoms in Dalton's system is the mass of hydrogen, the standard of the mass of atoms in SI is exactly 1/12 of the mass of 12C. This value is called the atomic mass unit (amu) and is equal to $1.6605402 \times 10^{-27}$ kg and D (Dalton) was assigned as its symbol. The atomic weight of an atom is defined as the ratio of the average amu of the element with its natural isotope distribution *vs.* 1/12 of amu of 12C.

Chapter 14

Magnetic Field, Diamagnetic, Solenoid

MAGNETIC PROPERTIES OF SOLIDS

Materials may be classified by their response to externally applied magnetic fields as *diamagnetic, paramagnetic,* or *ferromagnetic*. These magnetic responses differ greatly in strength.

Diamagnetism is a property of all materials and opposes applied magnetic fields, but is very weak. Paramagnetism, when present, is stronger than diamagnetism and produces magnetization in the direction of the applied field, and proportional to the applied field. Ferromagnetic effects are very large, producing magnetizations sometimes orders of magnitude greater than the applied field and as such are much larger than either diamagnetic or paramagnetic effects.

The magnetization of a material is expressed in terms of density of net *magnetic dipole moments* ? in the material. We define a vector quantity called the magnetization M by

$$M = \mu_{total}/V$$

Then the total *magnetic field* B in the material is given by

$$B = B_0 + \mu_0 M$$

where μ_0 is the magnetic *permeability* of space and B_0 is the externally applied magnetic field. When magnetic fields inside of materials are calculated using *Ampere's law* or the *Biot-Savart law,* then the μ_0 in those equations is typically replaced by just μ with the definition

$$\mu = K_m \mu_0$$

where K_m is called the *relative permeability*. If the material does not respond to the external magnetic field by producing any magnetization, then K_m = 1. Another commonly used magnetic quantity is the magnetic susceptibility which specifies how much the relative permeability differs from one.

Magnetic susceptibility $\div_m = K_m - 1$

For paramagnetic and diamagnetic materials the relative permeability is very close to 1 and the magnetic susceptibility very close to zero. For ferromagnetic materials, these quantities may be very large.

Another way to deal with the magnetic fields which arise from magnetization of materials is to introduce a quantity called *magnetic field strength H*. It can be defined by the relationship

$$H = B_0/\mu_0 = B/\mu_0 - M$$

and has the value of unambiguously designating the driving magnetic influence from external currents in a material, independent of the material's magnetic response. The relationship for B above can be written in the equivalent form

$$B = \mu_0(H + M)$$

H and M will have the same units, amperes/meter.

Ferromagnetic materials will undergo a small mechanical change when magnetic fields are applied, either expanding or contracting slightly. This effect is called *magnetostriction*.

Diamagnetism: The orbital motion of electrons creates tiny atomic current loops, which produce magnetic fields. When an external magnetic field is applied to a material, these current loops will tend to align in such a way as to oppose the applied field.

This may be viewed as an atomic version of *Lenz's law*: induced magnetic fields tend to oppose the change which created them. Materials in which this effect is the only magnetic response are called diamagnetic. All materials are inherently diamagnetic, but if the atoms have some net magnetic moment as in *paramagnetic* materials, or if there is *long-range ordering* of atomic magnetic moments as in *ferromagnetic* materials, these stronger effects are always dominant. Diamagnetism is the residual magnetic behaviour when materials are neither paramagnetic nor ferromagnetic.

Any conductor will show a strong diamagnetic effect in the presence of changing magnetic fields because circulating currents will be generated in the conductor to oppose the magnetic field changes. A *superconductor* will be a perfect diamagnet since there is no resistance to the forming of the current loops.

Paramagnetism: Some materials exhibit a magnetization which is proportional to the applied magnetic field in which the material is placed. These materials are said to be paramagnetic and follow Curie's law:

$$M = C\left(\frac{B}{T}\right) \quad \begin{matrix} M = \text{magnetization} & B = \text{magnetic field} \\ C = Curi's\ \text{constant} & T = \text{Temperaturein Kelvins} \end{matrix}$$

All atoms have inherent sources of magnetism because *electron spin* contributes a *magnetic moment* and electron orbits act as current loops which produce a magnetic field. In most materials the magnetic moments of the electrons cancel, but in materials which are classified as paramagnetic, the cancelation is incomplete.

Ferromagnetism: Iron, nickel, cobalt and some of the rare earths (gadolinium, dysprosium) exhibit a unique magnetic behaviour which is called ferromagnetism because iron (ferrum in Latin) is the most common and most dramatic example. Samarium and neodymium in alloys with cobalt have been used to fabricate very strong *rare-earth magnets.*

Ferromagnetic materials exhibit a *long-range ordering* phenomenon at the atomic level which causes the unpaired electron spins to line up parallel with each other in a region called a *domain.* Within the domain, the magnetic field is intense, but in a bulk sample the material will usually be unmagnetized because the many domains will themselves be randomly oriented with respect to one another.

Ferromagnetism manifests itself in the fact that a small externally imposed *magnetic field,* say from a *solenoid,* can cause the magnetic domains to line up with each other and the material is said to be magnetized.

The driving magnetic field will then be increased by a large factor which is usually expressed as a *relative permeability* for the material. There are many practical *applications* of ferromagnetic materials, such as the *electromagnet.*

Ferromagnets will tend to stay magnetized to some extent after being subjected to an external magnetic field. This tendency to "remember their magnetic history" is called *hysteresis.* The fraction of the saturation magnetization which is retained when the driving field is removed is called the *remanence* of the material, and is an important factor in permanent magnets.

All ferromagnets have a maximum temperature where the ferromagnetic property disappears as a result of thermal agitation. This temperature is called the *Curie temperature.*

Ferromagntic materials will respond mechanically to an impressed magnetic field, changing length slightly in the direction of the applied field. This property, called *magnetostriction,* leads to the familiar hum of transformers as they respond mechanically to 60 Hz AC voltages.

Long Range Order in Ferromagnets: The long range order which creates *magnetic domains* in *ferromagnetic* materials arises from a quantum mechanical interaction at the atomic level. This interaction is remarkable in that it locks the magnetic moments of neighboring atoms into a rigid parallel order over a large number of atoms in spite of the thermal agitation which tends to randomize any atomic-level order. Sizes of domains range from a 0.1 mm to a few mm.

When an external magnetic field is applied, the domains already aligned in the direction of this field grow at the expense of their neighbors.

If all the spins were aligned in a piece of iron, the field would be about 2.1 Tesla. A magnetic field of about 1 T can be produced in annealed iron with an external field of about 0.0002 T, a multiplication of the external field by a factor of 5000! For a given ferromagnetic material the long range order abruptly disappears at a certain temperature which is called the *Curie temperature* for the material. The Curie temperature of iron is about 1043 K.

The Curie Temperature: For a given *ferromagnetic* material the *long range order* abruptly disappears at a certain temperature which is called the Curie temperature for the material.

The Curie temperature of iron is about 1043 K. The Curie temperature gives an idea of the amount of energy takes to break up the long-range ordering in the material. At 1043 K the *thermal energy* is about 0.135 eV compared to about 0.04 eV at room temperature.

Magnetic Domains: The *microscopic ordering* of electron spins characteristic of *ferromagnetic* materials leads to the formation of regions of magnetic alignment called domains.

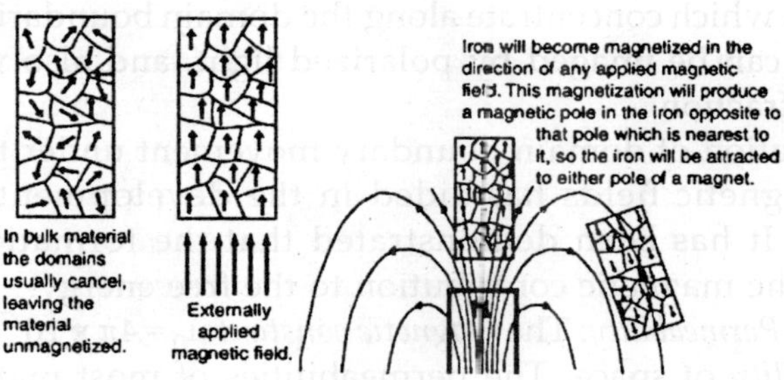

The main implication of the domains is that there is already a high degree of magnetization in ferromagnetic materials within individual domains, but that in the absence of external magnetic fields those domains are randomly oriented.

A modest applied magnetic field can cause a larger degree of alignment of the magnetic moments with the external field, giving a large multiplication of the applied field.

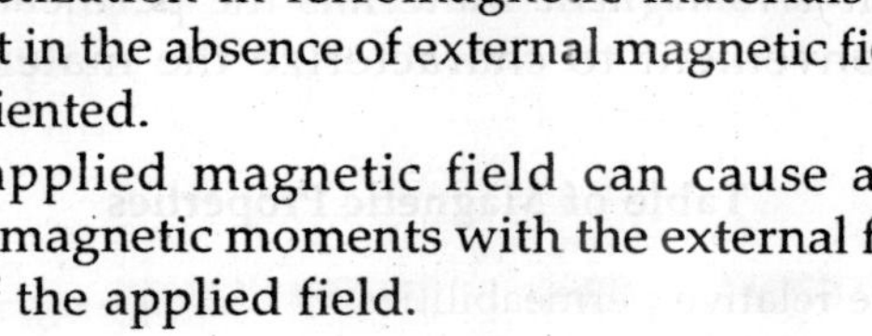

These illustrations of domains are conceptual only and not meant to give an accurate scale of the size or shape of domains. The microscopic evidence about magnetization indicates that the net magnetization of ferromagnetic materials in response to an external magnetic field may actually occur more by the growth of the domains parallel to the applied field at the expense of other domains rather than the reorientation of the domains themselves as implied in the sketch.

Some of the more direct evidence we have about domains comes from imaging of domains in single crystals of ferromagnetic materials. The sketches above are after Young and are adapted from magnified images of domain boundaries in single crystals of nickel.

They suggest that the effect of external magnetic fields is to cause the domain boundaries to shift in favour of those domains which are parallel to the applied field. It is not clear how this applies to bulk magnetic materials which are polycrystalline.

Keep in mind the fact that the internal magnetic fields which come from the *long range ordering* of the electron spins are much stronger, sometimes hundreds of times stronger, than the external magnetic fields required to produce these changes in domain alignment. The effective multiplication of the external field which can be achieved by the alignment of the domains is often expressed in terms of the *relative permeability*.

Domains may be made visible with the use of magnetic colloidal suspensions which concentrate along the domain boundaries. The domain boundaries can be imaged by polarized light, and also with the use of electron diffraction.

Observation of domain boundary movement under the influence of applied magnetic fields has aided in the development of theoretical treatments. It has been demonstrated that the formation of domains minimizes the magnetic contribution to the free energy.

Relative Permeability: The *magnetic constant* $\mu_0 = 4\pi \times 10^{-7}$ T m/A is called the *permeability* of space. The permeabilities of most materials are very close to μ_0 since most materials will be classified as either *paramagnetic* or *diamagnetic*. But in *ferromagnetic* materials the permeability may be very large and it is convenient to characterize the materials by a relative permeability.

Table of Magnetic Properties

Some representative relative permeabilities:

Material	Relative permeability	
magnetic iron	200	
nickel	100	at a magnetic
permalloy (78.5% nickel, 21.5% iron)	8,000	flux density of 0.002 W/m2
mumetal (75% nickel, 2% chromium, 5% copper, 18% iron)	20,000	

When ferromagnetic materials are used in applications like an *iron-core solenoid*, the relative permeability gives you an idea of the kind of multiplication of the applied magnetic field that can be achieved by having the ferromagnetic core present. So for an ordinary iron core you might expect a magnification of about 200 compared to the magnetic field produced by the solenoid current with just an air core. This statement has exceptions and limits, since you do reach a saturation magnetization of the iron core quickly, as illustrated in the discussion of *hysteresis*.

Magnetic Dipole Moment: From the expression for the *torque on a current loop*, the characteristics of the current loop are summarized in its magnetic moment

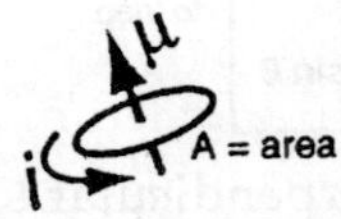

The magnetic moment can be considered to be a vector quantity with direction perpendicular to the current loop in the right-hand-rule direction. The torque is given by

$$\tau = \mu \cdot B$$

As seen in the geometry of a current loop, this torque tends to line up the magnetic moment with the *magnetic field* B, so this represents its lowest energy configuration. The *potential energy* associated with the magnetic moment is so that the difference in energy between aligned and anti-aligned is

$$U(\theta) = -\mu.B$$

These relationships for a finite current loop extend to the magnetic dipoles of *electron orbits* and to the intrinsic magnetic moment associated with *electron spin*. Also important are *nuclear magnetic moments*.

$$\Delta U = 2\mu.B$$

Torque on a Current Loop: The torque on a current-carrying coil, as in a *DC motor*, can be related to the characteristics of the coil by the "*magnetic moment*" or "magnetic dipole moment". The *torque* exerted by the *magnetic force* (including both sides of the coil) is given by

$$\tau = BILW \sin\theta$$

The coil characteristics can be grouped as called

$$\mu = IA (or\ \mu = NIA\ for\ n\ loops)$$

the magnetic moment of the loop, and the torque written as

$$\tau = \mu B \sin\theta$$

The direction of the magnetic moment is perpendicular to the current loop in the right-hand-rule direction, the direction of the normal to the loop in the illustration.

Considering torque as a *vector quantity*, this can be written as the *vector product*

$$\tau = \mu \times B$$

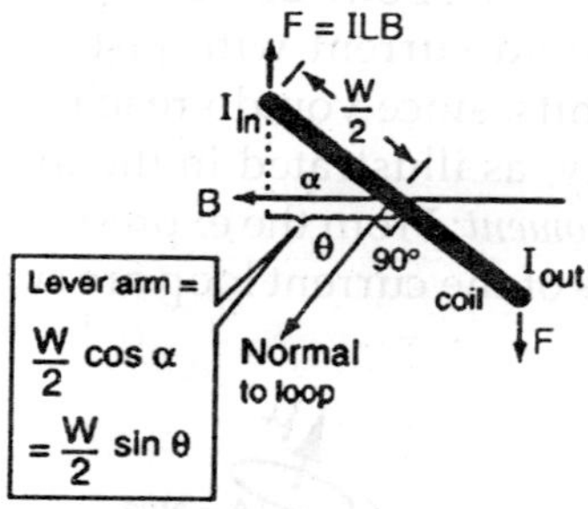

Since this torque acts perpendicular to the magnetic moment, then it can cause the magnetic moment to precess around the magnetic field at a characteristic frequency called the *Larmor frequency*.

If you exerted the necessary torque to overcome the magnetic torque and rotate the loop from angle zero to 180 degrees, you would do an amount of *rotational work* given by the *integral*

$$W = -\int_0^{\pi} \tau d\theta = -\int_0^{\pi} \mu B \sin\theta d0 = -\mu B \cos\theta |_0^{\pi} = 2\mu B$$

The position where the magnetic moment is opposite to the magnetic field is said to have a higher *magnetic potential energy*.

Chapter 15

Electricity and Light

OBJECTIVES

This teaching guide is designed to introduce the concepts of waves and seismic waves that propagate within the Earth, and to provide ideas and suggestions for how to teach about seismic waves.

The guide provides information on the types and properties of seismic waves and instructions for using some simple materials – especially the slinky – to effectively demonstrate seismic wave characteristics and wave propagation. Most of the activities described in the guide are useful both as demonstrations for the teacher and as exploratory activities for students.

With several regular metal slinkys, and the modified slinky demonstrations described in this teaching guide, one can involve an entire class in observation of the demonstrations and experimenting with the slinkys in small groups.

For activities that involve several people, such as the 5-slinky and human wave demonstrations, it is convenient to repeat the demonstrations with different groups of students so that each person will have the opportunity to observe the demonstration and to participate in it.

WAVES

Waves consist of a disturbance in materials (media), that carry energy and propagate. However, the material that the wave propagates in generally does not move with the wave. The movement of the material is generally confined to small motions, called particle motion, of the material as the wave passes.

After the wave has passed, the material usually looks just like it did before the wave, and, is in the same location as before the wave. (Near the source of a strong disturbance, such as a large explosion or earthquake, the wave-generated deformation can be large enough to cause permanent deformation which will be visible as cracks, fault offsets, and displacements of the ground after the disturbance has passed.) A source of energy creates the initial disturbance (or continuously generates a disturbance) and the resulting waves propagate (travel) out from the disturbance. Because there is finite energy in a confined or short-duration disturbance, the waves generated by such a source will spread out during propagation and become smaller (attenuate) with distance away from the source or with time after the initial source, and thus, will eventually die out.

Waves are often represented mathematically and in graphs as sine waves (or combinations or sums of sine waves) as shown in figure. The vertical axis on this plot represents the temporary motion (such as displacement amplitude A) of the propagating wave at a given time or location as the wave passes. The characteristics or properties of the wave – amplitude, wavelength, peaks, etc. – are illustrated in Figure 1.

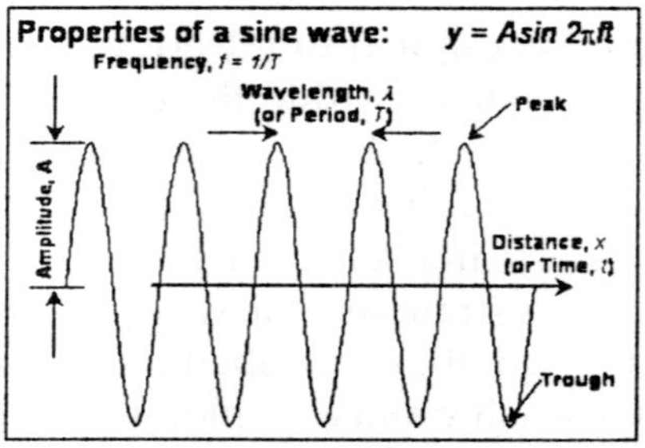

Figure: Properties of a sine wave. The horizontal axis displays time or distance and the vertical axis displays amplitude as a function of time or distance. Amplitude can represent any type or measure of motion for any direction. Seismic wave motion is commonly displayed with a similar plot, and sometimes the wave itself looks very similar to the sine wave shown here.

Additional information on the properties of waves can be found in Bolt (1993, p. 29-30) or Rutherford and Bachmeyer (1995). Sometimes the actual wave looks very much like the representation in the graph (Fig. 1). Examples are water waves and a type of seismic surface waves called Rayleigh waves. Commonly, propagating waves are of relatively short duration and look similar to truncated sine waves whose amplitudes vary with time.

Waves generated by a short duration disturbance in a small area, such as from an earthquake or a quarry blast, spread outward from the source as a single or a series of wavefronts. The wavefronts at successive times, and corresponding raypaths that show the direction of propagation of the waves, are illustrated in the figure.

A good model for illustrating wave motion of this type is water waves from a pebble dropped in a still pond or pool. The disturbance caused by the pebble hitting the surface of the water generates waves that propagate outward in expanding, circular wavefronts. Because there is more energy from dropping a larger pebble, the resulting waves will be larger (and probably of different wavelength).

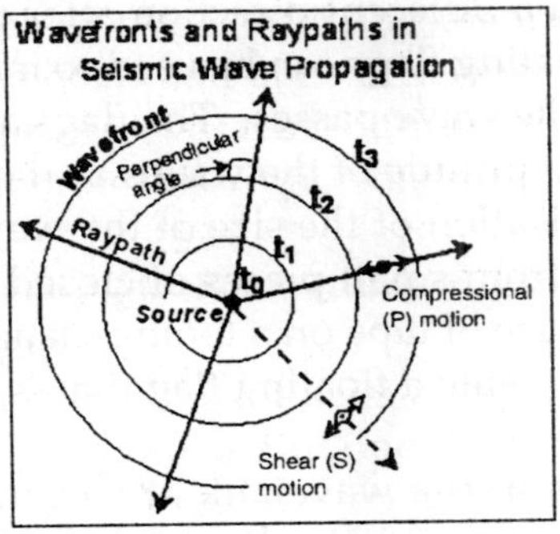

Fig. Wavefronts and Raypaths for a Seismic Wave Propagating from a Source.

Three positions (successive times) of the expanding wavefront are shown. Particle motions for P (compressional) and S (shear) waves are also shown. Raypaths are perpendicular to the wavefronts and indicate the direction of propagation of the wave.

P waves travel faster than S waves so there will be separate wavefront representations for the P and S waves. If the physical properties of the material through which the waves are propagating are constant, the wavefronts will be circular (or spherical in three-dimensions). If the physical properties vary in the model, the wavefronts will be more complex shapes.

WAVES IN WATER EXPERIMENT

Experiments with water waves in a small wave tank, consisting of a rectangular plastic storage box (about 60 to 120 cm long by 40 cm wide by 15 cm high; the exact size isn't important); a plastic storage box designed to fit under a bed, available at discount department stores such as K-Mart, Wal-Mart and Target, with about 5 cm of water in it, can easily illustrate the common properties of water waves.

Drops of water from an eyedropper or small spherical objects (a table tennis ball, a golf ball, or a small rubber ball works well) dropped onto the surface of the water are convenient sources.

With the wave tank one can demonstrate that: the size (amplitude) of the waves is related to the energy of the source (controlled by the mass and drop height); the waves expand outward (propagate) in circular wavefronts; the wave height decreases and eventually dies out with distance away from the source (or with time after the source) because of spreading out of the wave energy over a larger and larger area (or volume); the waves have a speed (velocity) of propagation that can be measured by placing marks every 10 cm on the bottom of the tank and timing the wave with a stopwatch; the waves reflect off the sides of the tank and continue propagating in a different direction after reflection.

About 3-4, small floating flags can be used to more effectively see the motion of the water as the wave passes. The flags are particularly useful for noting the relative amplitude of the waves as the degree of shaking of the flags is a visible indication of the size of the wave.

Flags can be made from small pieces of closed cell foam, styrofoam or cork. Place a small piece of tape on a toothpick and stick the toothpick into the foam or cork to create a floating flag that is sensitive to waves in the water.

By placing the flags in the wavetank at various distances from the source, one can easily observe the time of arrival and the relative amplitudes of the waves. A glass cake pan can be used as small wave tank on an overhead projector.

Waves generated by dropping drops of water from an eyedropper into about 2 cm of water in the tank will be visible on the screen projected from the overhead projector.

The velocity of propagation, attenuation of wave energy, and reflection of waves are important concepts for understanding seismic wave propagation, so experimenting with these wave properties in the water tank is a very useful exercise. Additional suggestions for experiments with water waves are contained in Zubrowski (1994).

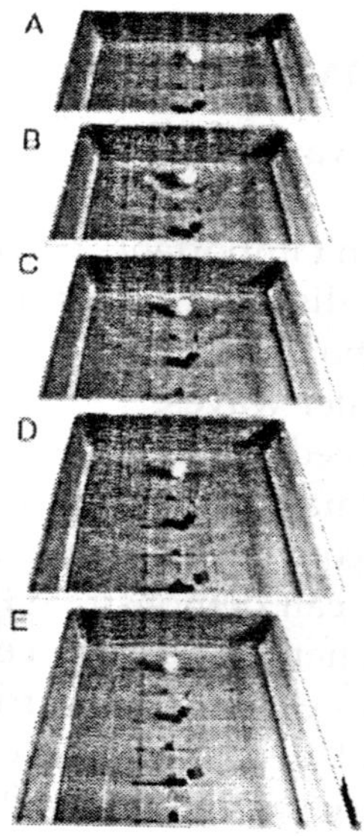

Fig. Large Wave Tank using a Plastic Storage Container (75 liter "Under Bed" Container).

Distances in cm can be marked on the bottom of the container for convenient measurement of wave velocity from the distance and travel time. Small floating flags are useful for identifying the time and relative amplitude of propagating water waves.

The flags are made from small floats with a toothpick and piece of tape attached. The floating flags are very sensitive to wave motion in the wave tank. In this sequence of photos, taken every one-half second, one can see the waves propagating outward from the source. Lines drawn on the bottom of the tank are spaced at 10 cm.

A. Source (a table tennis ball dropped from about 40 cm height) is just ready to hit the water surface.
B. (~ 0.5 s after the source) Water waves have propagated from the source to about 10 cm distance. A circular expanding wavefront is visible and the wave height is larger than in the later photos.
C. (~ 1.0 s) The wavefront has propagated to about 20 cm distance and the waves have decreased in amplitude.
D. (~ 1.5 s) The wavefront has propagated to about 30 cm distance.
E. (~ 2 s) The wavefront has propagated to about 40 cm distance and the wave amplitudes have decreased so much that they are difficult to see in this photo.

The small floating flags are sensitive detectors (similar to seismometers) of the waves. Because the waves have traveled about 40 cm in 2 seconds, the velocity of propagation is about 20 cm/s or 0.2 m/s.

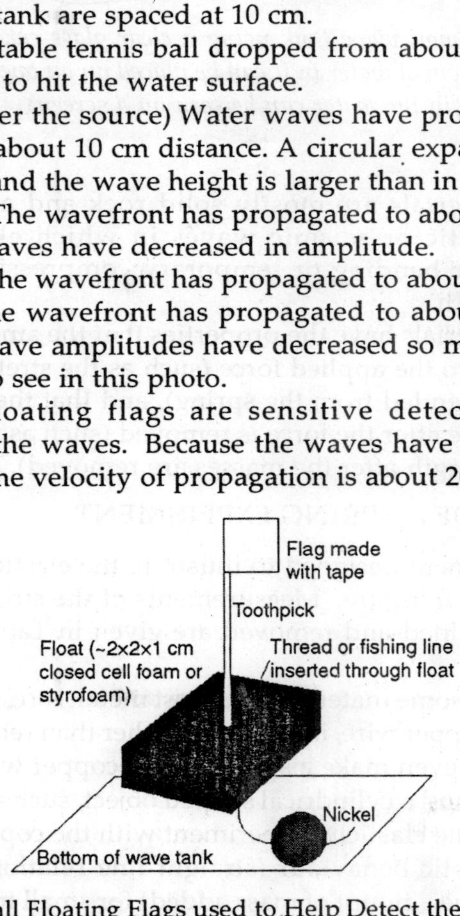

Fig. Small Floating Flags used to Help Detect the Motion of Water Waves in a Wave Tank.

The flags can be "anchored" using a length of thread and a nickel for a weight. The flags are sensitive to the motion caused by relatively small waves.

By noting the degree of shaking of the flags, one can identify (approximately) large, medium and small (barely detectible) wave action.

Be sure the water surface is calm (and the flags still) before initiating a water wave experiment.

Figure: 5. Small wave tank using a clear glass cake or baking dish. The dish with about 2 cm of water in it can be placed on an overhead projector so that wave propagation in the water can be seen on a screen.

ELASTICITY

Earth materials are mostly solid rock and are *elastic,* and thus propagate elastic or seismic waves in which elastic disturbances (deformation or bending or temporary compression of rocks) travel through the Earth.

Elastic materials have the properties that the amount of deformation is proportional to the applied force (such as the stretching of a spring as masses are suspended from the spring), and that the material returns to its original shape after the force is removed (such as the spring returning to its original length after the masses are removed).

ELASTICITY OF A SPRING EXPERIMENT

An experiment designed to illustrate the elastic characteristics of a spring is shown in figure. Measurements of the stretching of the spring as masses are added and removed are given in Table 1 and graphed in figure.

In contrast, some materials like most metals, are *ductile.* For example, if one bends a copper wire, it stays bent rather than returning to its original shape. One can even make a spring out of copper wire by wrapping the wire tightly around a cylindrical shaped object, such as a cardboard tube.

Repeating the elasticity experiment with the copper wire spring will result in an elastic behaviour (straight line relationships between the stretching and the amount of mass added) for small mass, and permanent deformation (stretching) for larger masses (the spring will not return to its original length as masses are removed).

Data for the copper wire spring experiment can be tabulated, as in Table 1, and graphed, as in figure.

Because this experiment produces a very different stretching versus added mass curve, as compared to the regular spring example, it is useful to ask the students before the experiment what they expect the results to look like and then compare the actual results with their predictions.

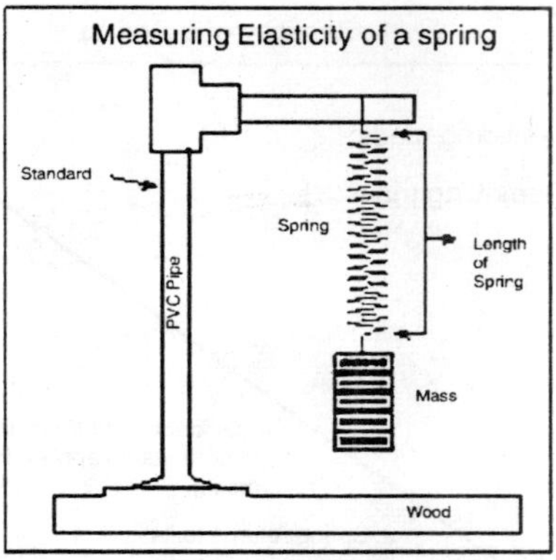

Fig. Schematic Diagram Illustrating Mass and Spring Apparatus for Measuring the Eelasticity (Stretching of the Spring as Mass is Added) of a Spring.

Table. Observations of Extension of a Spring Upon Adding and Removing Masses

Added Mass (g)	*Spring Extension (cm)* (adding masses)*	*Spring Extension (cm)* (removing masses)*
0	0.0	0.3
100	3.7	3.6
200	7.7	7.5
300	11.4	11.4
400	15.3	15.1

*Spring extension (stretching) is the length of the spring with mass attached minus the length of the spring with no mass. Different springs have different spring constants (elasticity) and will display different amounts of stretching for a given amount of added mass. Many springs are tightly-wound and require a small amount of force (suspended mass) to begin extending. For these springs, one should use the difference between this initial suspended mass and the total mass as the "added mass."

Measurements were made as mass was added and removed. The stretching of the spring is defined as the spring length (with added mass) minus the original length. Note that the extension versus mass data define a straight line. The slope of this line is related to the spring constant or elasticity of the spring. The two properties that characterize simple elasticity are described in the figure.

Seismic Waves: Unlike waves in water which are confined to a region very near to the water's surface, seismic waves also propagate through the Earth's interior.

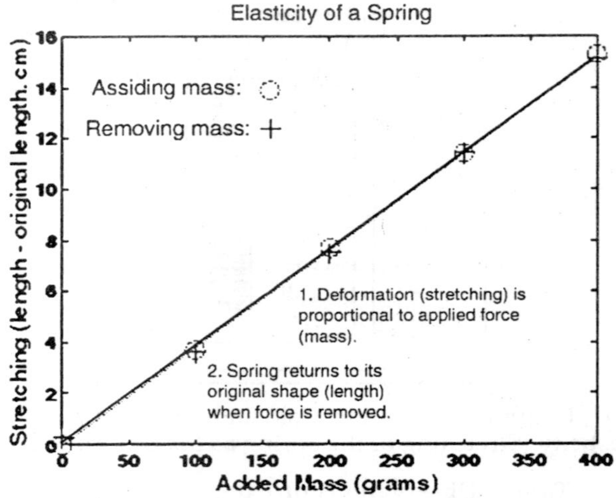

Fig. Graph Showing Measurements of Elasticity of a Spring.

Because of the elastic properties of Earth materials (rocks) and the presence of the Earth's surface, four main types of seismic waves propagate within the Earth. Compressional (P) and Shear (S) waves propagate through the Earth's interior and are known as body waves Love and Rayleigh waves propagate primarily at and near the Earth's surface and are called surface waves. Wave propagation and particle motion characteristics for the P, S, Rayleigh and Love waves are illustrated in figures. Further information and characteristics of these four kinds of seismic waves are given in table.

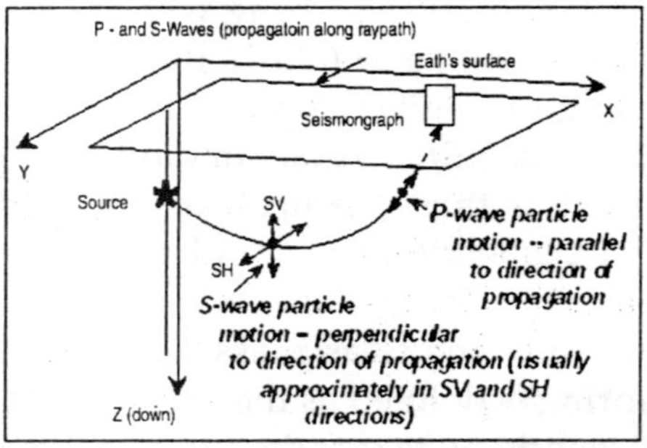

Fig. Raypath from the Source to a Particular Location on the Surface for P- and S- Wave Propagation in a Material with Variable Velocity.

P and S particle motion are shown. Because major boundaries between different rock types within the Earth are normally approximately parallel to the Earth's surface, S-wave particle motion is commonly in the SV (perpendicular to the raypath and vertical) and SH perpendicular to the raypath and horizontal) directions.

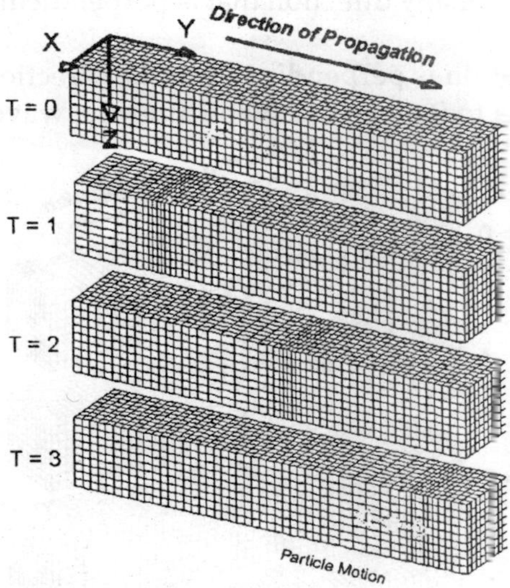

Perspective view of elastic P-wave propagation through a grid representing a volume of material. The directions X and Y are parallel to the Earth's surface and the Z direction is depth. T = 0 through T = 3 indicate successive times. The disturbance that is propagated is a compression (grid lines are closer together) followed by a dilatation or extension (grid lines are farther apart). The particle motion is in the direction of propagation. The material returns to its original shape after the wave has passed.

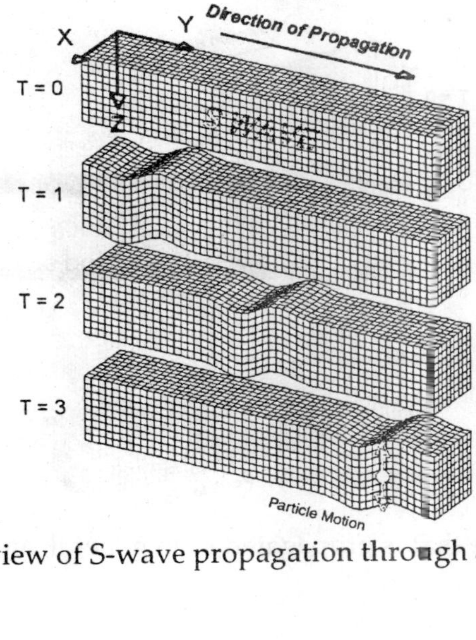

Perspective view of S-wave propagation through a grid representing a volume of elastic material. The disturbance that is propagated is an up motion followed by a down motion (the shear motion could also be directed horizontally or any direction that is perpendicular to the direction of propagation).

The particle motion is perpendicular to the direction of propagation. The material returns to its original shape after the wave has passed.

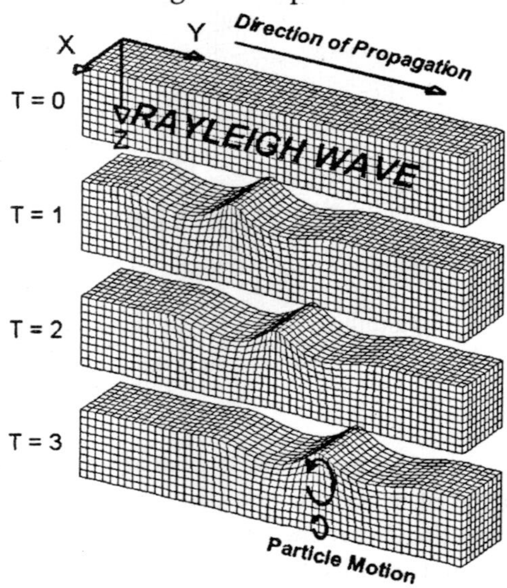

Perspective view of Rayleigh-wave propagation through a grid representing a volume of elastic material. Rayleigh waves are surface waves.

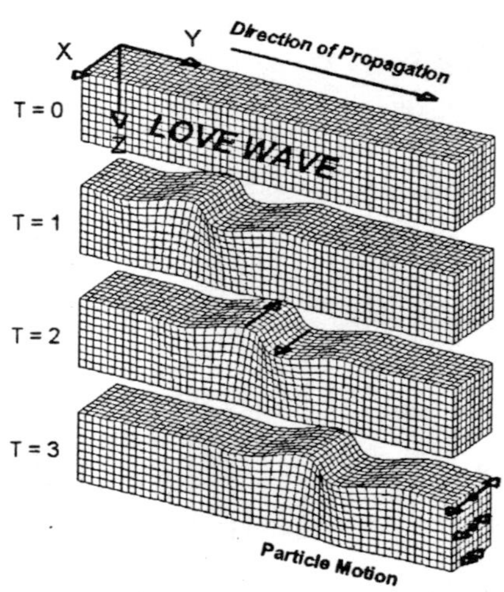

The disturbance that is propagated is, in general, an elliptical motion which consists of both vertical (shear; perpendicular to the direction of propagation but in the plane of the raypath) and horizontal (compression; in the direction of propagation) particle motion.

The amplitudes of the Rayleigh wave motion decrease with distance away from the surface. The material returns to its original shape after the wave has passed.

Perspective view of Love-wave propagation through a grid representing a volume of elastic material. Love waves are surface waves. The disturbance that is propagated is horizontal and perpendicular to the direction of propagation.

The amplitudes of the Love wave motion decrease with distance away from the surface. The material returns to its original shape after the wave has passed.

Table. Seismic Waves

Type (and names)	*Particle Motion*	*Typical Velocity*	*Other Characteristics*
P, Compressional, Primary, Longitudinal	Alternating compressions ("pushes") and dilations ("pulls") which are directed in the same direction as the wave is propagating (along the ray path); and therefore, perpendicular to the wavefront	VP ~ 5 - 7 km/s in typical Earth's crust; >~ 8 km/s in Earth's mantle and core; 1.5 km/s in water; 0.3 km/s in air	P motion travels fastest in materials, so the P-wave is the first-arriving energy on a seismogram. Generally smaller and higher frequency than the S and Surface-waves. P waves in a liquid or gas are pressure waves, including sound waves.
S, Shear, Secondary, Transverse	Alternating transverse motions (perpendicular to the direction of propagation, and the raypath); commonly polarized such that particle motion is in vertical or	VS ~ 3 - 4 km/s in typical Earth's crust; >~ 4.5 km/s in Earth's mantle; ~ 2.5-3.0 km/s in (solid) inner core	S-waves do not travel through fluids, so do not exist in Earth's outer core (inferred to be primarily liquid iron) or in air or water or molten rock (magma). S waves travel slower than P waves in a

	horizontal planes		solid and, therefore, arrive after the P wave.
L, Love, Surface waves, Long waves	Transverse horizontal motion, perpendicular to the direction of propagation and generally parallel to the Earth's surface	VL ~ 2.0 - 4.5 km/s in the Earth depending on frequency of the propagating wave	Love waves exist because of the Earth's surface. They are largest at the surface and decrease in amplitude with depth. Love waves are dispersive, that is, the wave velocity is dependent on frequency, with low frequencies normally propagating at higher velocity. Depth of penetration of the Love waves is also dependent on frequency, with lower frequencies penetrating to greater depth.
R, Rayleigh, Surface waves, Long waves, Ground roll	Motion is both in the direction of propagation and perpendicular (in a vertical plane), and "phased" so that the motion is generally elliptical either prograde or retrograde	VR ~ 2.0 - 4.5 km/s in the Earth depending on frequency of the propagating wave	Rayleigh waves are also dispersive and the amplitudes generally decrease with depth in the Earth. Appearance and particle motion are similar to water waves.

SLINKY DEMONSTRATIONS OF P AND S WAVES

The P and S waves have distinctive particle motions and travel at different speeds. P and S waves can be demonstrated effectively with a slinky.

For the P or compressional wave, have two people hold the ends of the slinky about 3-4 meters apart. One person should cup his or her hand over the end (the last 3-4 coils) of the slinky and, when the slinky is nearly at rest, hit that hand with the fist of the other hand.

The compressional disturbance that is transmitted to the slinky will propagate along the slinky to the other person. Note that the motion of each coil is either compressional or extensional with the movement parallel to the direction of propagation.

Because the other person is holding the slinky firmly, the P wave will reflect at that end and travel back along the slinky. The propagation and reflection will continue until the wave energy dies out. The propagation of the P wave by the slinky is illustrated in figure.

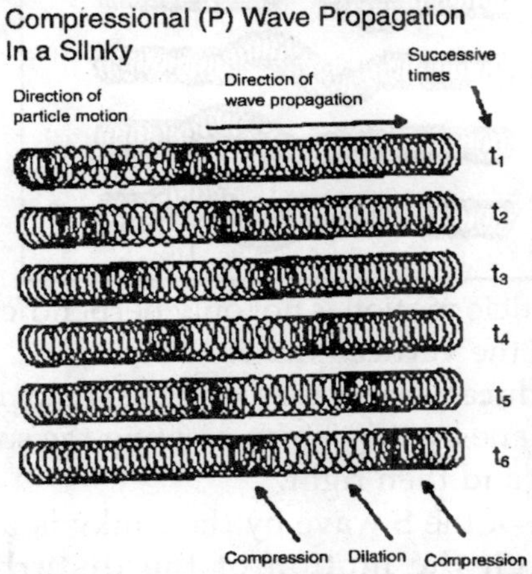

Compressional (P) wave propagation in a slinky. A disturbance at one end results in a compression of the coils followed by dilation (extension), and then another compression.

With time (successive times are shown by the diagrams of the slinky at times t_1 through t_6), the disturbance propagates along the slinky. After the energy passes, the coils of the slinky return to their original, undisturbed position. The direction of particle motion is in the direction of propagation.

Shear (S) wave propagation in a slinky. A disturbance at one end results in an up motion of the coils followed by a down motion of the coils. With time (successive times are shown by the diagrams of the slinky at times t1 through t6), the disturbance propagates along the slinky. After the energy passes, the coils of the slinky return to their original, undisturbed position. The direction of particle motion is perpendicular (for example, up and down or side to side) to the direction of propagation.

Demonstrating the S or Shear wave is performed in a similar fashion except that the person who creates the shear disturbance does so by moving his or her hand quickly up and then down. This motion generates a motion of the coils that is perpendicular to the direction of propagation, which is along the slinky.

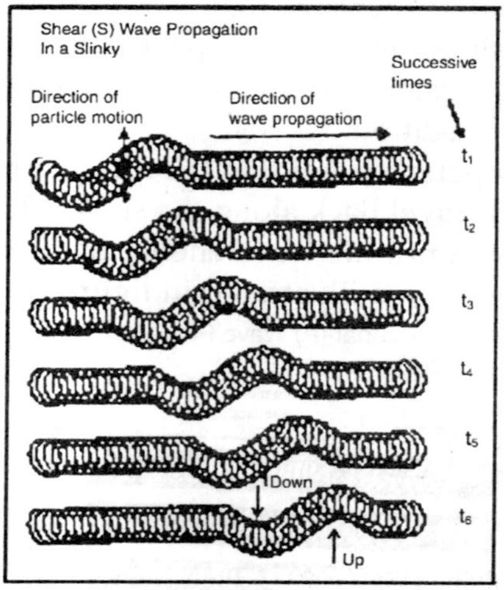

Note that the particle motion is not only perpendicular to the direction of motion but also in the vertical plane.

One can also produce Shear waves with the slinky in which the motion is in the horizontal plane by the person creating the source moving his or her hand quickly left and then right.

The propagation of the S wave by the slinky is illustrated in figure. Notice that, although the motion of the disturbance was purely perpendicular to the direction of propagation (no motion in the disturbing source was directed along the slinky), the disturbance still propagates away from the source, along the slinky.

The reason for this phenomenon (a good challenge question for students) is because the material is elastic and the individual coils are connected (like the individual particles of a solid) and thus transmit their motion to the adjacent coils. As this process continues, the shear disturbance travels along the entire slinky (elastic medium).

P and S waves can also be generated in the slinky by an additional method that reinforces the concept of elasticity and the elastic rebound theory which explains the generation of earthquakes by plate tectonic movements.

In this method, for the P wave, one person should slowly gather a few of the end coils of the slinky into his or her hand. This process stores elastic energy in the coils of the slinky that are compressed (as compared to the other coils in the stretched slinky) similar to the storage of elastic energy in rocks adjacent to a fault that are deformed by plate motions prior to slip along a fault plane in the elastic rebound process. When a few coils have been compressed, release them suddenly (holding on to the end coil of the slinky) and a compressional wave disturbance will propagate along the slinky.

This method helps illustrate the concept of the elastic properties of the slinky and the storage of energy in the elastic rebound process. However, the compressional wave that it generates is not as simple or visible as the wave produced by using a blow of one's fist, so it is suggested that this method be demonstrated after the previously-described method using the fist.

Similarly, using this "elastic rebound" method for the S waves, one person holding the end of the stretched slinky should use their other hand to grab one of the coils about 10-12 coils away from the end of the slinky. Slowly pull on this coil in a direction perpendicular to the direction defined by the stretched slinky.

This process applies a shearing displacement to this end of the slinky and stores elastic energy (strain) in the slinky similar to the storage of strain energy in rocks adjacent to a fault or plate boundary by plate tectonic movements. After the coil has been displaced about 10 cm or so, release it suddenly (similar to the sudden slip along a fault plane in the elastic rebound process) and an S wave disturbance will propagate along the slinky away from the source.

ILLUSTRATION OF ENERGY CARRIED BY THE WAVES

The fact that the seismic waves (P or S) that propagate along the slinky transmit energy can be illustrated effectively by using a slinky in which one end (the end opposite the source) has a small wood block attached. The wood block has a cardboard model building attached to it as shown in figures.

As P or S wave energy that propagates along the slinky is transmitted to the wood block, the building vibrates. This model is a good demonstration of what happens when a seismic wave in the Earth reaches the surface and causes vibrations that are transmitted to houses and other buildings. By generating P, S-vertical and S-horizontal waves that transmit vibration to the model building, one can even observe differences in the reaction of the building to the different directions of motion of the propagating wave.

Diagram showing how to attach a slinky to the edge of a small wood block with screws and washers. A cardboard "building" is attached with tape to the wood block. The model building is made from manila folder (or similar) material and is taped together. The diagram is a side view of the slinky, wood block and model building. When seismic waves are propagated along the slinky, the vibrations of the cardboard building show that the wave energy is carried by the slinky and is transmitted to the model building.

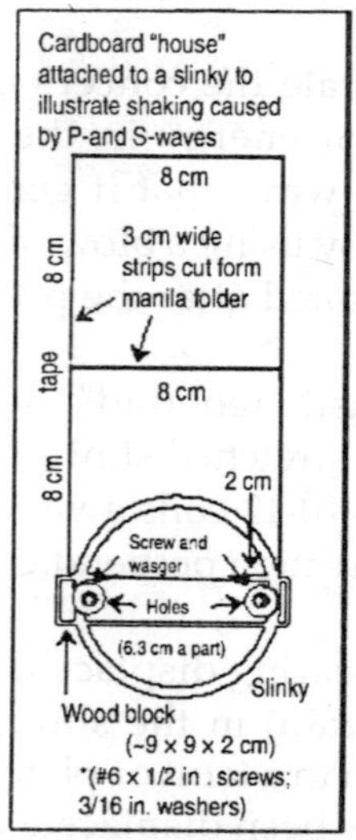

Photograph of slinky attached to a small wood block and cardboard "house." The slinky could also be attached to the bottom to the wood block to illustrate P- and S-waves propagating upwards and shaking the building.

WAVE PROPAGATION IN ALL DIRECTIONS

An additional demonstration with P and S waves can be performed with the 5-slinky model. By attaching 5 slinkys to a wood block as shown in figure, 5 people can hold the ends of the 5 slinkys (stretched out in different directions to about 3-4 m each). One person holds the wood block and can generate P or S waves (or even a combination of both) by hitting the wood block with a closed fist or causing the block to move quickly up and then down or left and then right.

The purpose of this demonstration is to show that the waves propagate in all directions in the Earth from the source (not just in the direction of a single slinky).

Attaching an additional slinky (with small pieces of plastic electrical tape) to one of the five slinkys attached to the wood block makes one slinky into a double length slinky which can be stretched out to 6-8 m. For one of the other four slinkys, have the person holding it collapse about half of the coils and hold them in his or her hands, forming a half slinky, stretched out about 1½ - 2 m.

Now when a source is created at the wood block, one can see that the waves take different amounts of time to travel the different distances to the ends of the various slinkys.

An effective way to demonstrate the different arrival times is to have the person holding each slinky call out the word "now" when the wave arrives at their location. The difference in arrival times for the different distances will be obvious from the sequence of the call of "now."

This variation in travel time is similar to what is observed for an earthquake whose waves travel to various seismograph stations that are different distances from the source (epicenter).

Although these two demonstrations with the five slinky model represent fairly simple concepts, we have found the demonstrations to be very effective with all age groups. In fact, the five slinky demonstrations are often identified as the "favourite" activities of participants.

Five Slinky P and S Wave Propagation Demonstration

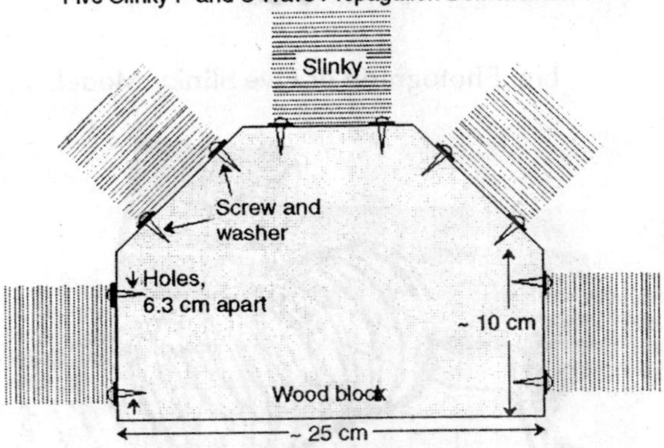

Diagram showing how five slinkys can be attached to the edge of a wood block. Photographs of the five slinky model are shown in figures. When the slinkys are stretched out to different positions (five people hold the end of one slinky each) and a P or S wave is generated at the wood block, the waves propagate out in all directions. The five slinky model can also be used to show that the travel times to different locations (such as seismograph stations) will be different. To demonstrate this effect, wrap a small piece of tape around a coil near the middle of the slinky for one of the slinkys.

Have the person holding that slinky compress all of the coils from the outer end to the coil with the tape so that only one half of the slinky is extended.

Also, attach an additional slinky, using plastic electrical tape, to the end of one of the slinkys. Have the person holding this double slinky stand farther away from the wood block so that the double slinky is fully extended.

When a P or S wave is generated at the wood block, the waves that travel along the slinky will arrive at the end of the half slinky first, then at approximately the same time at the three regular slinkys, and finally, last at the double length slinky. The difference in travel time will be very noticeable.

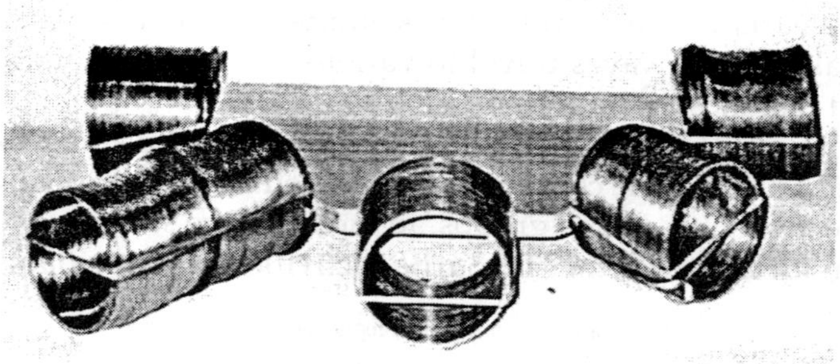

Fig. Photograph of Five Slinky Model.

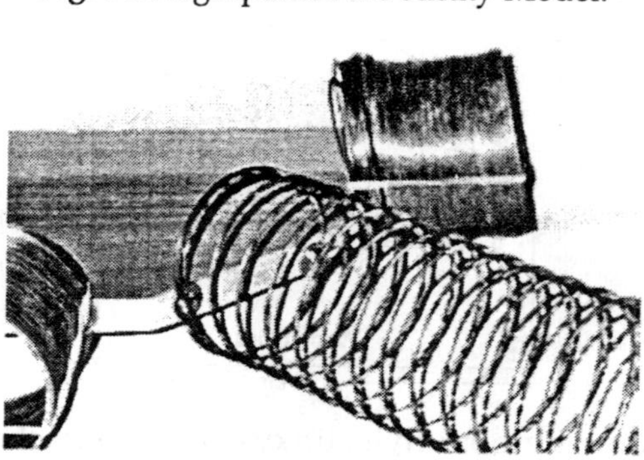

Fig. Close-up View of Five Slinky Model Showing Attachment of Slinky Using Screws and Washers.

HUMAN WAVE DEMONSTRATION - P AND S WAVES IN SOLIDS AND LIQUIDS

This demonstration involves a class or group in simulating seismic wave propagation in solids and liquids. If you have 20 or more people in the group, half can perform the demo while the other half watches; then switch places. The concepts that are involved in this demonstration are very dramatically illustrated to the participants. Once they have done the human wave activity, they should always remember the properties of P and S waves propagating in both solids and liquids. Have about 10 people stand at the front of the room, side by side, with their feet about shoulder width apart. Instruct the group to not be too rigid or too limp when pushed from the side.

They should give with the force that they will feel from the person next to them, but not fall over, and then return to their upright position. In other words, they should be "elastic". Have a "spotter" at the end of the line in case the last person begins to fall. (It is important to stress these instructions to the participants so that the demonstration will work effectively and so that participants do not fall over as the wave propagates down the line of people.)

To represent wave propagation in a solid, have each person put their arms over the shoulders of the person next to them ("chorus line style"; the "molecules" or "particles" of the solid are tightly bonded). Push on the person at the end of the line and the deformation (leaning to the side and then straightening up) will propagate down the line of people approximating a P wave.

Note that the propagation down the line took some time (there is a velocity for the wave propagation) and that although each person was briefly subjected to a deformation or disturbance, the individuals did not move from their original locations. Also, the motion of each person as the wave passed was in the direction of propagation and that, as the wave passed, the people moved closer together temporarily (compression), and then apart (dilation) to return to their original positions.

Human wave demonstration for the P wave in a solid. Instructor begins the P wave motion (in this case from left to right) in the line by pushing (compressing) on the first person in the line and then pulling the person back to an upright position.

For the S wave in a solid, make the first person at the end of the line bend forward at the waist and then stand up straight. The transverse or shear motion will propagate down the line of people. Again, the wave takes some time to propagate and each person ends up in the same location where they started even though a wave has passed.

Also, note that the shear motion of each particle is perpendicular to the direction of propagation. One of the observers can time the P and S wave propagation in the human wave using a stopwatch.

Because the shear wave motion is more complicated in the human wave, the S wave will have a slower velocity (greater travel time from source to the end of the line of people), similar to seismic waves in a solid.

Human wave demonstration for the S wave in a solid. The teacher starts the S wave motion (in this case propagating from right to left) in the line by causing the first person in the line to bend forward at the waist and then stand up straight.

Next, to represent wave propagation in a liquid, have the people stand shoulder-to-shoulder, without their arms around each other. Push on the shoulder of the end person and a P wave will propagate down the line.

The P wave will have the same characteristics in the liquid as described previously for the solid. Now, make the person at the end of the line bend forward at the waist - a transverse or shear disturbance.

However, because the "molecules" of the liquid are more loosely bound, the shearing motion will not propagate through the liquid (along the line of people). The disturbance does not propagate to the next person because the liquid does not support the shearing motion.

(Compare pressing your hand down on the surface of a solid such as a table top and on the surface of water and moving your hand parallel to the surface.

There will be considerable resistance to moving your hand on the solid. One could even push the entire table horizontally by this shearing motion. However, there will be virtually no resistance to moving your hand along the surface of the water.)

Only the first person in the line - the one that is bent over at the waist - should move because the people are not connected. If the next person bends, "sympathetically", not because of the wave propagating, ask that person if he or she felt, rather than just saw, the wave disturbance, then repeat the demonstration for S waves in a liquid.

VELOCITY OF WAVE PROPAGATION EXPERIMENT

Developing an understanding of seismic wave propagation and of the velocity of propagation of seismic waves in the Earth is aided by making measurements of wave speed and comparing velocities in different materials. For waves that travel an approximately straight line, the velocity of propagation is simply the distance traveled (given in meters or kilometers, for example) divided by the time of travel (or "travel time") in seconds.

Using a stopwatch for measuring time and a meter stick or metric tape for measuring distance, determine the wave velocity of water waves in the wave tank, the P wave in the slinky, and the P wave in the human wave experiment. Also, determine the velocity of sound in air using the following method. On a playfield, measure out a distance of about 100 meters. Have one person with a stopwatch stand at one end of the 100 meter line.

Have another person with a metal garbage can and a stick stand 100 meters away from the person with the stopwatch. Have the person with the stick hit the garbage can so that the instant of contact of the stick with the garbage can is visible from a distance.

The person with the stopwatch should start the stopwatch when the stick strikes the can and stop it when the sound generated by the stick hitting the can is heard. The measured speed of sound will be the distance divided by the travel time measured on the stopwatch.

(This measurement assumes that the speed of light is infinite - a reasonable approximation as the actual speed of light is about 3 × 108 or 300 million meters/second, much faster than the speed of sound, and that the reaction time for the person operating the stopwatch is about the same for starting the watch when the can is struck and for stopping the watch when the sound is heard.

The measurement should be repeated a few times to obtain an estimate of how accurately the measurement can be made. An average of the time measurements can be used to calculate the sound speed. The difference between the arrival of a light wave and associated sound is commonly used to determine how far away a lightning strike is in a thunderstorm. For example, because the speed of sound in air is about 330 m/s, if the difference in time between seeing a lightning strike and hearing the associated thunder is 3 seconds, the lightning is 1 km away; similarly, 6 s for 2 km away, etc.)

Make a list of the wave velocities (in m/s) for the water waves, slinky, human wave, and sound wave in air. (Measured wave speeds should be approximately 0.25-0.5 m/s for water waves in a wave tank, 2 m/s for the compressional human wave, 3 m/s for P waves in the slinky, and 330 m/s for sound waves in air.)

Compare these wave velocities with the compressional wave velocity in the Earth which varies from about 1000 m/s for unconsolidated materials near the Earth's surface to about 14,000 m/s in the Earth's lowermost mantle (1 to 14 km/s). The seismic velocity in solid rocks in the Earth is controlled by rock composition (chemistry), and pressure and temperature conditions, and is found to be approximately proportional to rock density (density = mass/unit volume; higher density rocks generally have larger elastic constants resulting in higher seismic velocity).

ATTENUATION OF WAVES

To demonstrate the property of anelasticity, a model with two slinkys can be constructed. Anelasticity is the absorption of energy during propagation which causes waves to attenuate in addition to the attenuation caused by the energy spreading out (for example, like the spreading of water waves created by dropping a pebble into a pond).

This effect is an important concept in evaluating earthquake hazards and comparing the hazards in two locations such as the western United States and the eastern United States. Seismic waves propagate very efficiently in the eastern United States resulting in damage over a wide area from a large earthquake.

In contrast, waves propagating in the western United States are attenuated by absorption of energy to a much greater degree. Thus, although earthquakes of a given size occur much more often in the western US, the earthquake hazard in the eastern US is significant because the area of damage (for equivalent earthquakes) is larger in the eastern US.

Attach two slinkys to a 30 cm long piece of "one by two" (1" by 2" wood trim) using washers and screws similar to that shown in figure. For one of the slinkys, place a strip of foam (about a 3 m strip of ½" thick foam 7 cm wide; the 3 m strip can be pieced together from 2-3 shorter pieces; tape or sew together the strips) in the extended coils. The foam should fit snugly within the coils and will absorb wave energy that is propagated along the slinky.

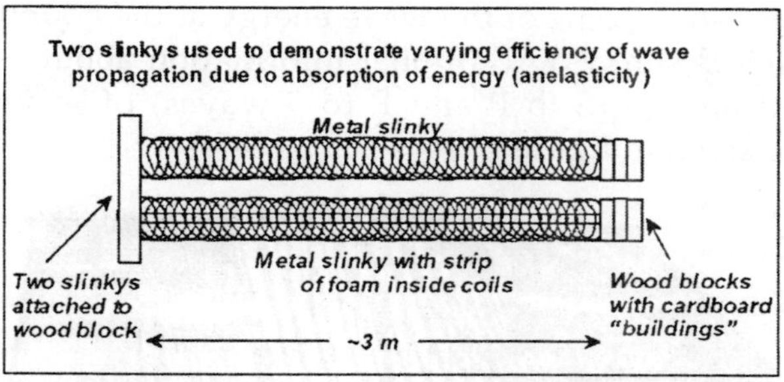

Schematic diagram illustrating the construction of a model using two slinkys (one with a foam strip inside the coils of the slinky) that can be used to demonstrate the concept of absorption of energy by material during propagation (anelasticity). A photo of the slinky model for demonstrating absorption is shown in figure.

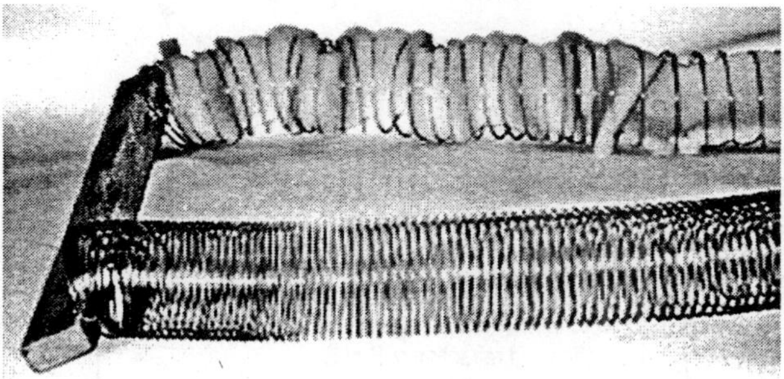

Slinky model used to demonstrate attenuation of waves by absorption of energy.

Small wood blocks and model buildings can be attached to the free ends of both the regular slinky and the foam-lined slinky so that the wave energy that reaches the end of the slinky will be more visible for comparison. Extend each slinky about 3 m and cause P or S waves to be generated simultaneously in both slinkys by hitting the wood block (for P waves) or quickly moving the wood block vertically or horizontally (for S waves).

The regular slinky will propagate the waves very efficiently while the slinky with the foam will strongly attenuate the wave energy (by absorbing some of the energy) as it propagates.

REFLECTION OF WAVES

The reflection of wave energy at a boundary between two types of materials can be demonstrated with slinkys by attaching a regular metal slinky to a plastic slinky. The attachment can be made with small pieces of plastic electrical tape. Generating P or S waves in the metal slinky will result in reflection of some of the wave energy at the boundary between the two stretched slinkys. Additional information about reflection and conversion of energy (S to P and P to S waves) of seismic waves at boundaries is given in figure

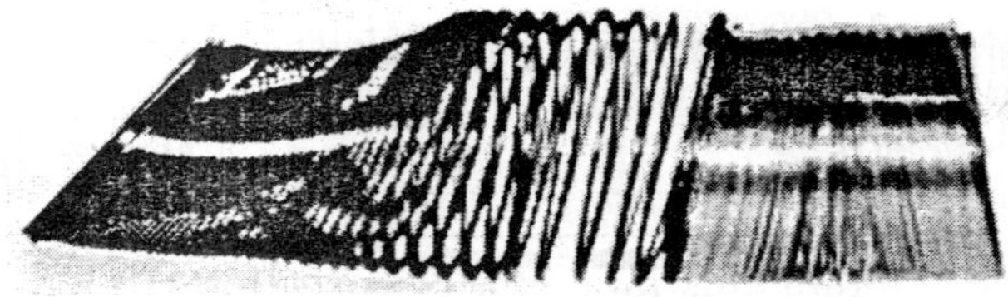

Two slinkys - a metal slinky and a plastic slinky - attached together using small pieces of electrical tape. The two slinkys can be used to illustrate reflection and transmission of wave energy at a boundary between elastic materials with differing elastic properties.

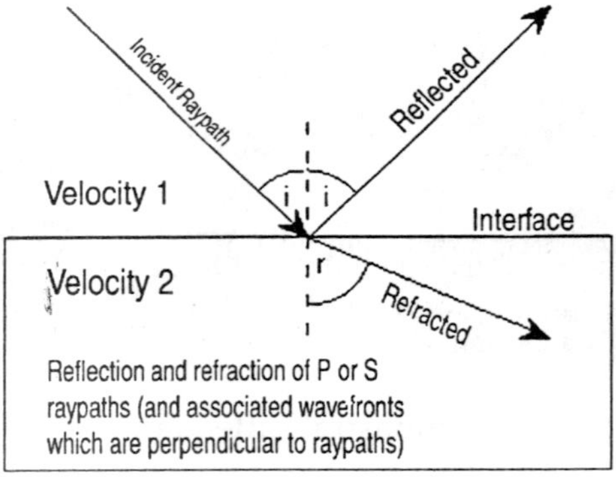

Reflection and refraction (transmission) of seismic waves (P or S waves) at an interface separating two different materials. Some energy is reflected and some is transmitted. The effect can be illustrated with two slinkys - one metal and one plastic - taped together. Waves traveling along one slinky are partially reflected at the boundary between the two types of slinkys.

For seismic waves in the Earth, an incident P or S wave also results in converted S and P energy in both the upper material (reflected) and the lower material (transmitted). If the angle of approach of the wave (measured by the angle of incidence, i, or the associated raypath) is not zero, the resulting transmitted P and S raypaths will be bent or refracted at the boundary. A similar change in angle is also evident for the reflected, converted waves.

SURFACE WAVES

The Love wave is easy to demonstrate with a slinky or a double length slinky. Stretch the slinky out on the floor or on a tabletop and have one person at each end hold on to the end of the slinky. Generate the Love wave motion by quickly moving one end of the slinky to the left and then to the right.

The horizontal shearing motion will propagate along the slinky. Below the surface, the Love wave motion is the same except that the amplitudes decrease with depth.

Using the slinky for the Rayleigh wave is much more difficult. With a regular slinky suspended between two people, one person can generate the motion of the Rayleigh wave by rapidly moving his or her hand in a circular or elliptical motion.

The motion should be up, back (toward the person generating the motion), down, and then forward (away from the person), coming back to the original location and forming an ellipse or circle with the motion of the hand.

This complex pattern will propagate along the slinky but will look very similar to an S wave. Excellent illustrations of the wave motion of Love and Rayleigh waves can also be found in Bolt. Rayleigh wave motion also decreases with depth below the surface. Further details on the characteristics and propagation of Love and Rayleigh waves can be found in Bolt.

OSCILLATIONS OF THE WHOLE EARTH

When a very large earthquake occurs, long period surface wave energy penetrates deep within the Earth and propagates all around the globe. At particular points around the globe, the timing between this wave energy (which just keeps circling the globe for many hours or even days before dying out or attenuating) results in constructive and destructive interference of the wave energy.

The resulting oscillations at certain frequencies are called normal modes or free oscillations of the Earth and are vibrations of the whole Earth. Further information about Earth normal modes, the phenomenon of standing waves, and diagrams illustrating the modes of vibration of the whole Earth are available in Bolt.

SEISMIC WAVES IN THE EARTH

Seismic body waves (P and S waves) travel through the interior of the Earth. Because confining pressure increases with depth in the Earth, the velocity of seismic waves generally increases with depth causing raypaths of body waves to be curved.

Because the interior structure of the Earth is complex and because there are four types of seismic waves (including dispersive surface waves), seismograms, which record ground motion from seismic waves propagating outward from an earthquake (or other) source, are often complicated and have long (several minutes or more) duration. An effective computer simulation that illustrates wave propagation in the Earth is the programme Seismic Waves by Alan Jones.

Using this programme, which shows waves propagating through the interior of the Earth in speeded-up-real-time, one can view the spreading out of wavefronts, P, S, and surface waves traveling at different velocities, wave reflection and P-to-S and S-to-P wave conversion. The programme also displays actual seismograms that contain arrivals for these wave types and phases. Exploring wave propagation through the Earth with the Seismic Waves programme is an excellent follow-up activity to the seismic wave activities presented in this teaching guide.

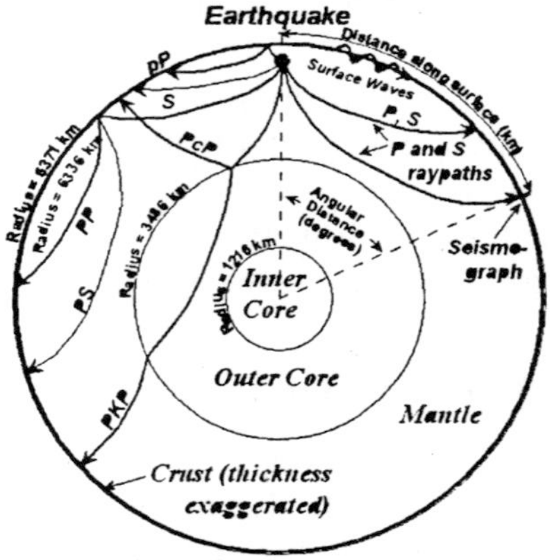

Figure. Cross section through the Earth showing important layers and representative raypaths of seismic body waves. Direct P and S raypaths (phases), including a reflection (PP and pP), converted phase (PS), and a phase that travels through both the mantle and the core (PKP). P raypaths are shown by heavy lines. S raypaths are indicated by light lines. Additional information about raypaths for seismic waves in the whole Earth and illustrations of representative raypaths are available in Bolt. Surface wave propagation (Rayleigh waves and Love waves) is schematically represented by the heavy wiggly line. Surface waves propagate away from the epicenter, primarily near the surface and the amplitudes of surface wave particle motion decrease with depth.

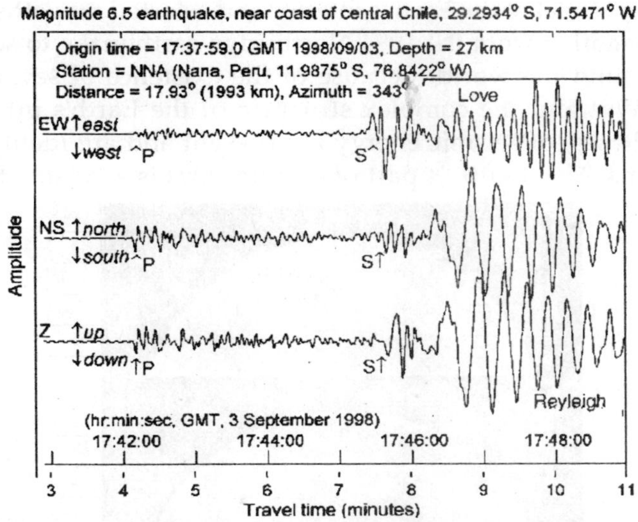

Seismograms recorded by a 3-component seismograph at Nana, Peru for an earthquake located near the coast of central Chile on September 3, 1998. The three seismograms record motion in the horizontal (east-west and north-south) and the vertical (Z) directions. P, S, Rayleigh and Love waves are identified on the record.

The S wave arrives significantly after the P-wave because S-wave velocity in rocks is lower than P wave velocity. Additional arrivals between the P and the S wave are P and S waves that have traveled more complicated paths (such as the pP and PP phases and P-to-S converted phases) from the earthquake location to the seismograph.

The surface waves arrive after the S waves because surface wave velocities in rocks are lower than the shear wave velocity. The surface waves extend over a long time interval because surface wave propagation is dispersive (the velocity of propagation is dependent on the frequency of the wave). This dispersive character can easily be seen in the Rayleigh wave on the vertical (Z) component seismogram in that the earliest Rayleigh wave energy has a longer period (lower frequency; see figure) than the later arriving waves. The seismic data that are displayed here were acquired from the IRIS using the WILBER programme from the IRIS Data Management Centre.

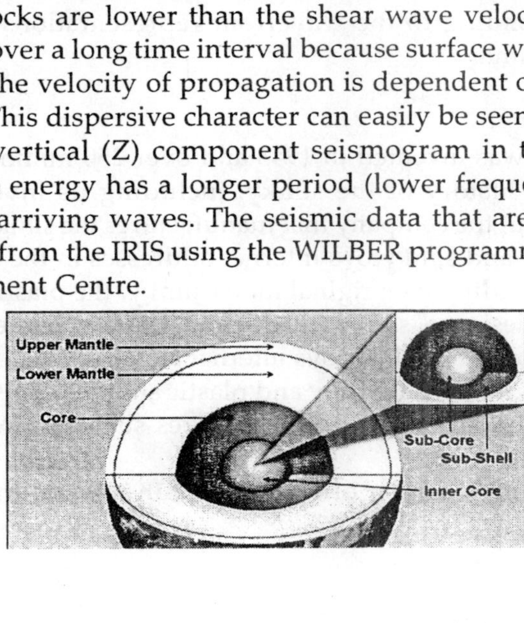

Figure: Model of the Earth's interior and selected raypaths for seismic wave propagation from the 1994 Northridge earthquake to seismograph stations around the world. Because of the existence of several types of seismic waves and the complex structure of the Earth's interior, many arrivals (phases) of seismic energy are present and are identified on the seismograms. This figure is part of a poster that is available from IRIS.

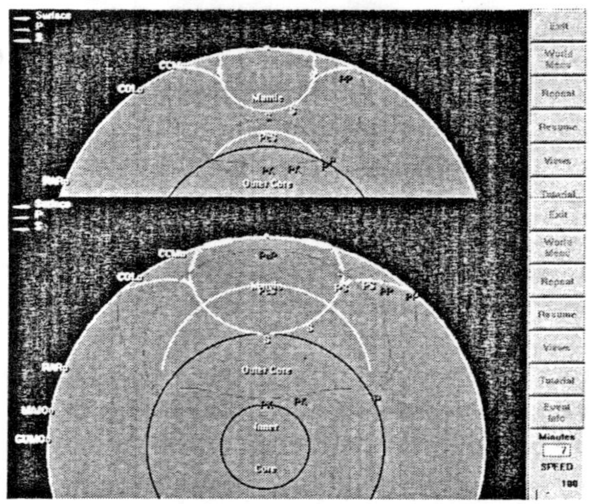

Partial screen views of the Seismic Waves computer programme. The upper image shows seismic wavefronts traveling through the Earth's interior five minutes after the earthquake.

The lower image shows the wavefronts 7 minutes after the earthquake. P waves are shown in red; S waves are shown in blue; and Surface waves are shown in yellow. Three and four letter labels on the Earth's surface show relative locations of seismograph stations that recorded seismic waves corresponding to the wavefront representations in the Seismic Waves programme.

THE SLINKY

The slinky was invented in 1943 and over 250 million of them have been sold. The history of the slinky, including its invention and the information about the company that manufactures the slinky, can be found at the discovery and slinkytoys internet addresses in the reference list.

Slinkys, including the original metal slinky, the plastic slinky, slinky junior, several "designer" slinkys made from a variety of metals, and slinky toys can be found at the slinkytoys internet address listed in the reference list. Both the original metal slinky and plastic slinkys are usually available (for about $2) at discount department stores such as K-Mart, Wal-Mart and Target.

A "long" slinky can be ordered, but can also be made by taping together two regular slinkys with small pieces of plastic electrical tape to make a double-length slinky. The original metal slinky is the most effective type of slinky for most of the wave propagation activities.

A double-length slinky is useful for illustrating Love wave motion (and the concept of standing waves) on the floor. The plastic slinky does not propagate compressional or shear waves as efficiently as the metal slinky, but is useful for illustrating reflection and transmission of wave energy at a boundary between two elastic materials with different properties.

For this demonstration, tape a metal slinky and a plastic slinky together by attaching the end coils with small pieces of plastic electrical tape.

Slinkys. A. Original metal slinky; B. Plastic slinky; C. Long slinky; D. Slinky junior.

Slinky lessons for teaching, including physics and wave activities, can be found at the Newtons (from the Newton's Apple PBS television programme), and the teaching tools and eecs.umich (National Engineers Week activities) internet addresses.

The National Engineers Week slinky activities are associated with a video ("Slinky Science Shindig") that is available from the eweek.org address.

Slinkys are educational and fun! It is useful to have many of them available for the activities described in this teachers guide and for your students to perform the activities, and to experiment and discover.

SUMMARY

The slinky in various forms provides an excellent model to demonstrate and investigate seismic wave characteristics and propagation. A summary of the slinky models and their uses in demonstrations and activities is given in table. Additional information on seismic waves, wave propagation, earthquakes and the interior of the Earth can be found. Illustrations of seismic wave propagation through the Earth and seismograms are contained in Bolt

Table: Slinky Types, Models and Demonstrations

Number	*Slinky Types/Models*	*Demonstrations*	*Figure Number*
1	Regular metal slinky	P and S waves	13, 14
2	Long slinky (or attach 2 regular slinkys together with plastic electrical tape)	Love wave on floor or tabletop	---
3	Slinky with cardboard "building"	Illustrate that P and S energy is propagating and causes the card-board building to shake as wave arrives. Differences in shaking can be seen for P and S motion	15, 16
4	Two slinkys (plastic and metal) attached with plastic electrical tape	Reflection and trans-mission of energy at a boundary between materials of different types (elastic properties or seismic velocities)	24, 25
5	Five slinkys attached to wood block	Show that waves propagate in all direc tions from source; that travel time is different to different dis-tances; and that wave vibration for P and S sources will be diffe-rent in different direc-tions from the source	17, 18, 19
6	Two slinkys attached to wood block, one with soft foam within the coils; cardboard buil-dings can be attached to the slinkys to help see the differences in shaking	Illustrate the concept of attenuation due to absorption of energy during propagation and and some materials propagate waves more efficiently than other materials	22, 23

Notes to the Teacher: The activities described in this teachers guide are designed for both classroom demonstration and inquiry-based activities for students. The elastic properties of a spring and the waves in a water tank activities are appropriate for student experiments.

Several of the slinky activities and the human wave demonstration actively involve students in the class. The slinky demos should also be performed by the students to increase their understanding of the wave propagation characteristics and concepts.

All of the activities provide opportunities for developing and practicing observation skills and experience with measuring and timing. These demonstrations and activities are suitable for inclusion in Earth science teaching at a variety of levels.

At higher grade levels, more complete treatment, increased emphasis on necessary vocabulary (defining the properties of elasticity, describing the characteristics of motion of the different types of seismic waves, etc.), involving quantitative measurements (measuring and graphing the elasticity of a spring, calculating the velocity of propagation of waves, etc.), and connection to related seismology and other Earth science lessons and activities are desirable.

The depth of investigation and the length of time devoted to the seismic wave activities will depend on the grade level and characteristics of the students, time available, and teacher preference. The teacher will also need to determine the "degree of constructivism" to be employed in the teaching strategy for these demonstrations and activities.

For example, several of the activities, such as the P- and S-wave propagation in a slinky, could be performed first as a demonstration and then by the students to develop further understanding by first-hand experience.

Alternatively, the teacher could choose to have a brief classroom discussion about P and S waves and then challenge the students to use the slinky to discover P- and S-wave propagation and determine the primary characteristics of the waves (property of propagation, material returns to original condition after the wave has passed, reflection of wave energy, particle motion) and the distinctive differences between P and S waves.

This student exploration would then be followed by a "final" demonstration to emphasize the key concepts and clarify any misunderstandings. The seismic wave activities should normally be included in an earthquake unit of an Earth science curriculum that also covers plate tectonics and the causes of earthquakes, Earth's interior structure, seismographs and seismograms, earthquake location methods, and earthquake hazards.

Appropriate assessment methods for the activities and science content presented here can include both written and oral responses by students and specific assessment activities.

The scope and depth of questioning will depend on class level, time devoted to the seismic wave activities and how much related Earth science material (such as studies of plate tectonics, earthquake statistics and hazards, Earth structure, etc.) has been covered. If student teams have completed the elasticity of a spring, waves in water, and/or velocity of propagation experiments, the teacher will have some written material from the results of the student's experiments that can be assessed.

Here are some suggested activities that can be used for authentic assessment:

1. Have students perform the P and S wave slinky demonstrations and describe their observations and the wave characteristics that they observe.
2. Have students repeat the elasticity experiment with a spring that has a different spring constant (a "weaker" or "stronger" spring). Have students predict what the graphed line will look like in comparison with their previous result, then perform the experiment and graph the results. Their predictions should be close to the actual results.
3. Ask students to predict what would happen if a rubber band was used instead of the spring in the elasticity experiment (they should describe the expected result in terms of the two main properties of elasticity).
4. Provide the students with a graph showing a hypothetical seismic wave. Have the students interpret the graph and identify principal wave characteristics (frequency, amplitude, peaks, etc.).
5. Provide the students with copies of unlabeled diagrams illustrating the four types of seismic waves (lower three grids of the perspective views of P, S, Rayleigh and Love wave propagation in figures; remove labels identifying wave type). Have the students identify which diagram corresponds to the four types of seismic waves and what characteristics of the wave motion allow them to identify the wave type.
6. Obtain the Seismic Waves computer programme and demonstrate to the students or provide the students with the opportunity to run the programme. Next, with the programme available on a monitor, start waves propagating (pause or restart as necessary) for one of the earthquakes and have the students watch the wavefront diagram (interior of Earth; it is convenient to set the display to view only the Earth cross section view; use

the "Options" menu and "Select View ..." to choose the cross section view) and have them answer the following questions. What are the approximate shapes of the initial (in the upper mantle near the source) P and S wavefronts?

Why are they shaped that way? From the P and S wavefronts, estimate the relative velocity of the S wave as compared to the P wave (for example, how much longer does it take for the S wave to arrive at a station or to travel from the source to the core-mantle boundary as compared to the P wave)?

How long does it take for the P wave to travel directly through the Earth to the opposite side of the Earth from the source? This distance (the diameter of the Earth) is about 12,742 km. From these measurements, what is the approximate average velocity for P waves in the Earth (in km/s)? Explain the new wavefronts that are generated when the P or S wave hits the core-mantle boundary.

Why is there no S wave that travels directly through the Earth to the other side? Can there be S waves in the inner core (the programme may not be of much help here except to visualize wavefronts that propagate in the Earth's core because no S wave phases from the inner core are displayed; not all seismic phases or raypaths are illustrated by the programme)?

How could S waves be generated so that they would travel through the inner core? Open the whole Earth view (surface of Earth with oceans and continents is visible). How are the patterns of the wavefronts that you see propagating and expanding from the source similar to the water waves in the wave tank experiment (or generated by a pebble dropped into a mud puddle or a pond)? How are they different?

Below are some suggested questions for a written assessment.

1. What is the source of energy for the generation of seismic waves; in the slinky demonstrations? in the Earth?
2. How could we demonstrate that energy is carried by the waves in the slinky demonstrations?
3. How can we determine the velocity of propagation of a wave?
4. Explain the property of elasticity.
5. Explain how the slow motions of the Earth's plates (like slowly deforming the stretched slinky in the P and S demonstrations alternative methods of wave generation) can produce a rapid release of energy (slip along a fault in an earthquake, release of stored elastic energy in the slinky) resulting in seismic waves propagating in the Earth (or the slinky).

The activities and the science content contained in this teachers guide have significant connections to the National Science Education Standards.

Table. "Seismic Waves and the Slinky: A Guide for Teachers" and the National Science Education

NSES Standard	How standard is addressed by Seismic Waves and the Slinky demonstrations, lessons and activities*
Science Teaching Standards	Many of the activities are inquiry-based (A, B) and provide opportunities for ongoing assessment (C).
Professional Development Standards	The guide for teachers provides opportunities and appropriate resource material for teachers to learn about an Earth science topic that is not likely to have been included in their previous educational experiences (A, C) and includes suggestions for effective teaching strategies (B).
Assessment Standards	Authentic assessment activities and questions for assessing achievement in learning key concepts are included (C).
Science Content Standards *- Unifying Concepts and Processes in Science* *- Science as Inquiry*	Activities provide experience with observation, evidence and explanation, and constancy, change and measurement. Activities provide opportunities for practice of inquiry and of fundamental science skills (Grades 5-8 and 9-12, A).
– Physical Science Standards	Activities explore properties and changes of properties in matter, motion and forces, transfer of energy (Grades 5-8, B). Activities explore structure and properties of matter, motions and forces, and interactions of energy and matter (Grades 9-12, B).
– Earth and Space Science	Activities explore structure of the Earth system (Grades 5-8, D). Activities relate to energy in the Earth system (Grades 9-12, D).
– Science in Personal and *Social Perspectives*	Activities explore natural hazards (Grades 5-8, F). Activities explore natural and human-induced hazards (Grades 9-12, F).

Science Education Programme Standards	Seismic wave activities are developmentally appropriate, interesting and relevant, and emphasize student understanding through inquiry, and are connected to other school subjects (B). Seismic wave activities provide practice with mathematics and analysis skills (C). Activities provide experience with a variety of materials and resources for experimentation and direct investigation of phenomena (D).
Science Education System Standards	Because only relatively simple and inexpensive resources are necessary to perform the seismic wave demonstrations and activities, they are easily accessible to all students.

*Letters in parentheses identify specific standards within the six areas (Science Teaching, Professional Development, Assessment, Science Content, Science Education Programs, and Science Education System Standards) of the NSES

REFRACTIVE INDEX

Nature provides a whole host of superior multifunctional structures that can be used as inspirational systems for the design and synthesis of new, technologically important materials and devices. We review here the exceptional optical performance of microlens arrays formed by light-sensitive brittlestars, their structural and compositional features, and advantageous properties.

We show that brittlestars form a nearly perfect optical device with micron-scale, lightweight, mechanically strong, aberration-free, birefringence-free, individually-addressed lenses, which offer a unique focusing effect, signal enhancement, intensity adjustment, angular selectivity, and photochromic activity.

We discuss first materials fabrication strategies that were inspired by the principles involved in the formation of echinoderm calcitic structures. The biomimetic synthetic microlens arrays could be potentially used as highly tunable optical elements for a wide variety of applications.

The increasing technological requirement for a new generation of optical devices with novel architectures, tunability and tailored properties, provides a potent stimulus to the academic study of optical systems in living organisms. Mimicking Nature's methods of biological manufacture is proving to be a major step forward in modern technology. In general, a successful bio-inspired engineering effort includes three key objectives:

(i) To search for smart biological solutions in the design, synthesis, and integration of complex materials and systems;

(ii) To learn about their composition, structure, properties, tunability and mechanisms of formation; and (iii) to identify new bio-inspired synthetic strategies and to apply this knowledge to the fabrication of novel, superior materials and devices.

We are interested in discovering natural optical systems, whose hierarchical architecture and hybrid character offer outstanding optical properties and enable multi-faceted roles for these units. Biology provides a multitude of varied, new paradigms for the development of adaptive optical networks.

Recently, we have shown that biologically formed optical systems are often unique in their ability to perform multiple functions - optical and structural. For example, the study of light-sensitive ophiuroids showed that the skeletal dorsal arm plates of some species not only afford protection, but also form highly efficient microlens arrays.

Other research has concentrated on the exceptional fibre-optical properties of siliceous spicules of certain hexactinellid sponges, whose primary function is structural (either skeletal or anchoring). This feature article describes the composition, design and optics of the natural ophiuroid microlenses, summarizes their advantageous materials properties, and shows our first, bioinspired synthetic efforts in the fabrication of new optical structures.

Structural and Optical Characterization of Biologically Formed Microlenses

Echinodermata, the group of animals including sea stars and sea urchins, show various levels of photosensitivity. An especially interesting example is presented by brittlestars (Ophiuroidea) in the genus Ophiocoma, which exhibit strong light responses.

It has been generally believed that many echinoderms have an extraocular sensitivity to light that is supported by unspecialized, 'diffuse', dermal receptors. The absence of specialized 'eyes' is not, however, readily reconciled with such reactions as the colour-change and negative phototactic behaviour observed in certain species. The variations in colour and escape mechanisms are most striking in a brittlestar Ophiocoma wendtii.

The individuals are a dark brown colour during the day, and are strikingly banded with gray and black during the night. It has been documented that O. wendtii is able to escape from predatory fish by hiding in coral crevices, and that it may be able to sense shelter at a distance. Recently, we suggested that these strong responses to light might be related to the presence of a characteristic lensar extension of the brittlestar skeleton, which may constitute an advanced photoreception system. Echinoderms use calcium carbonate for their skeleton construction.

Fig. The Same Individual of the Brittlestar O. Wendtii, Photographed during the Day (top) and During the Night (bottom).

Each skeletal element (ossicle) is a single crystal of calcite. Unlike biogenic calcite of other phyla or abiologically formed calcite crystals, echinoderm skeletons are composed of a unique, intricately shaped, three-dimensional, single-crystalline meshwork with smooth and curved surfaces lacking crystal faces (so-called, stereom).

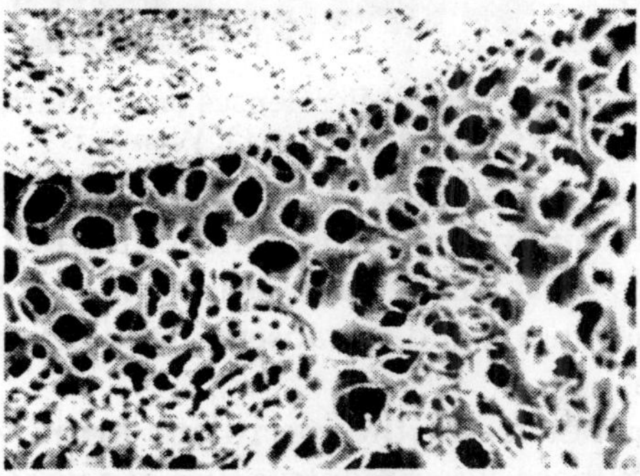

Scanning electron micrograph (SEM) of a typical stereom structure in echinoderm skeletons. The entire elaborate mesh is a single calcite crystal.

In the living animal, the calcite meshwork is filled with soft tissue (so-called, stroma), and the ossicles are attached to one another by connective tissue and muscle. Every joint of a brittlestar arm is composed of five major ossicles, including two lateral arm plates that support several spines and tentacle scales, one dorsal and one ventral arm plate. These surround the arm and enclose a large internal skeletal ossicle (vertebra) that is adapted for articulation. The analysis of the ultrastructure of these skeletal units showed that while the stereom in the vertebrae, spines, tentacle scales, and ventral arm plates generally share a typical design, the dorsal arm plates (DAPs) and portions of the lateral arm plates in certain species of Ophiocoma exhibit an unusual extension of the stereom - a thick, transparent, layer (ca. 40 mm thick) that covers the external surface of the plate.

It is composed of a close-set array of hemispherical calcitic structures (reaching 40-50 mm in diameter) with a characteristic double-lens design. The lenses are uniform in shape and appear as the scaled replicas of each other.

The ratio between the lens thickness (t) and the lens diameter (L) estimated for a statistically significant set of microlenses was:

$$t / L = 0.9 \pm 0.05$$

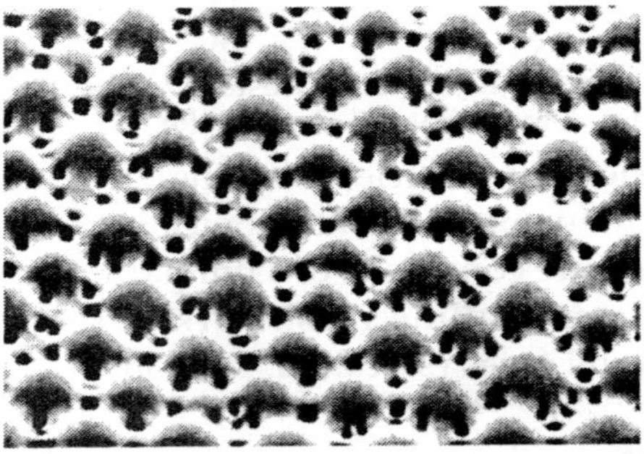

Fig. SEM of An Array of Microlenses on the Surface of the Dorsal Arm Plate in O. Wendtii.

The degree of development of the lensar layer seems to correlate with the photosensitivity among several ophiocomid brittlestars that were studied. The layer is particularly pronounced in a highly photosensitive species, Ophiocoma wendtii, and it is absent in the light-indifferent species, Ophiocoma pumila.

The optical properties of the microlens array were tested using photolithography. In these experiments, a film of a photosensitive material (positive photoresist) was illuminated through the isolated lensar layer. The illumination dose (I) was fixed slightly below the sensitivity level of the photoresist (I0). As a result, the developed photoresist film was only affected precisely where the illumination dose increased due to the focusing activity of the lenses.

The photoresist was placed at different distances h from the array. The photoresist films appeared to be selectively exposed under each microlens. Thus, the lensing effect was confirmed and mapped, and its quantitative characteristics (such as the position of the focal plane, d; the coefficient of the intensity enhancement by the microlenses, E; the angular selectivity, w; the size of the focused beam at the focal plane, a0; etc.) were experimentally determined from the values of h, L and the sizes of the spots in the photoresist, a, using basic equations of the geometric optics for a thick lens.

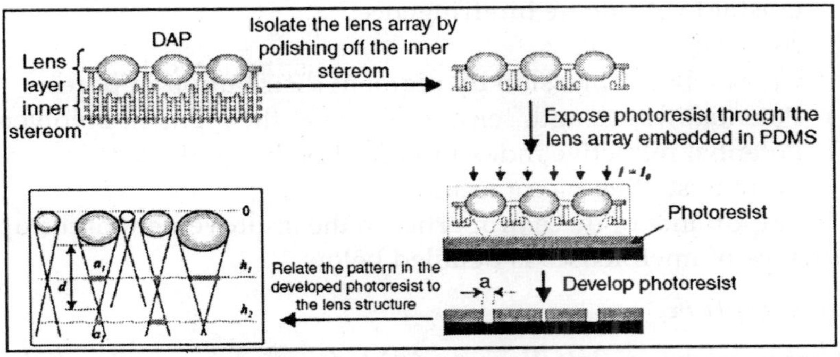

Fig. Schematic Presentation of the Lithographic Experiment, in which the Focusing Capabilities of the Lens Array were Studied.

Our analysis showed that the average position of the focal plane is located at the distance d = 4-7 mm below the microlenses. The size of the spot in the focal plane a0 is approximately 3 mm. The intensity of the incoming light is enhanced at the focal point by a factor of E#50. The angular selectivity w is about 10u. These results presented the first experimental evidence that the microlens arrays are effective optical elements capable of significantly enhancing and guiding the light inside the tissue. Calcitic microlenses of analogous structure were also reported in trilobite eyes.

Advantages of the Natural Microlens Design

The optical characteristics of the external DAP layer obtained in the imaging experiments showed that brittlestar microlenses are in many aspects superior to their man-made analogs.

This bio-optical system presents a spectacular example of Nature's ability to evolve sophisticated solutions even to complex technological problems.

We have identified a number of unique, advantageous properties and design strategies found in the brittlestar microlenses, which could be emulated in synthetic microlens arrays:

(a) High fill factor lens array with characteristic micron-scale porous structure.
(b) Individually-addressed crystalline lenses.

(c) Wide-range transmission adjustment using pigment rearrangement.
(d) Characteristic lens geometry that minimizes spherical aberration;
(e) Fracture toughness due to the formation of an inorganic- organic composite.
(f) Crystallographic orientation along the optical axis of the lens crystal to eliminate birefringence.
(g) Angular selectivity.
(h) Cross-talk suppression by pigment distribution in pores.
(i) Potential wavelength selectivity due to the pigment involvement.
(j) Potential refractive index modification by gradient of specialized in tracrystalline biomolecules.

The importance of these properties in the improvement of the optical performance of microlenses is detailed below.

Signal Reception

It has been shown that the remarkable focusing properties of the bio-microlenses are used by the brittlestars in their natural habitat for survival purposes. In a series of transmission electron microscopy (TEM) studies, Hendler and Byrne have identified neural bundles under the lenses, and suggested these to act as primary photoreceptors in the light signal detection.

This hypothesis was further confirmed in our lithographic experiments. The nerve bundles are positioned in the pores of the stereom at a distance of about 5 mm from the lens, which corresponds well to the location of the experimentally identified focal point d. Moreover, the diameter of the nerve fibers correlates with the size a0 of the focused beam in the focal plane.

Therefore, the arrays of lenses appear to form a sophisticated optical device, in which each component (a single microlens) is individually-addressed by the detector positioned at the focal point of the lens. Owing to the observed angular selectivity of these lenses and their different orientation, such a device could act as a compound eye that provides a wide field of view due to the non-planar arrangement of the lenses, which can be tuned on demand when the arm moves. The ways in which the signals detected by each receptor are processed and integrated remains unknown, and are worthy of further investigation.

Transmission Tunability

One of the most noticeable reactions to light in certain species of ophiocomid brittlestars is diurnal colour change. The change from a dark colour during the day to a light colour at night cannot be explained by camouflage, as has been shown experimentally. Hendler and Byrne have shown that the channels in the stereom contain chromatophore cells filled with pigment granules.

They proved that the diurnal colour change is caused by the migration of these cells. The light microscopic sections of DAPs prepared from night and day samples showed that during the day, pigment-filled processes of the chromatophore beneath the epidermis cover the lenses. During the night the chromatophores withdraw into the stereom channels surrounding the lenses, deeper within the DAP.

As a result, the intensity of the light reaching the receptors is regulated by chromatophores depending on the illumination conditions; a function performed by an iris in a human eye. A similar process is utilized in so-called 'transition' sunglasses. The behaviour of O. wendtii suggests that the transmitted light signal is tuned to match the sensitivity level of the neural bundle and to optimize the reception.

The intensity-adjustment function of the chromatophore cells as well as the involvement of the sub-lens nerve bundles in photoreception (as opposed to a 'diffuse' dermal reception) were corroborated by neurophysiological data. Cobb and Hendler studied the reaction of O. wendtii to light stimuli by direct recording from the nerve cord in brittlestar arms.

Nervous response of light-adapted, dark brown pigmented brittlestars has been monitored after the chemical disruption (by bleaching) of successive layers of the tissue. The removal of the epidermal layer only (bleaching for less than 10 s) did not result in any change in the response compared to untreated, control specimens.

Thus, 'diffuse' photoreception by epithelial cells was ruled out. The removal of the pigment layer (bleaching for ca. 30 s) caused an approximately ten-fold increase in the response to light 'ON', confirming the involvement of pigment in photoreception and its 'sunglasses' function. The disruption of the sub-lens layer (bleaching for w45 s) resulted in no photic responses, consistent with the suggestion that the sub-lens neural bundles are the primary photoreceptors).

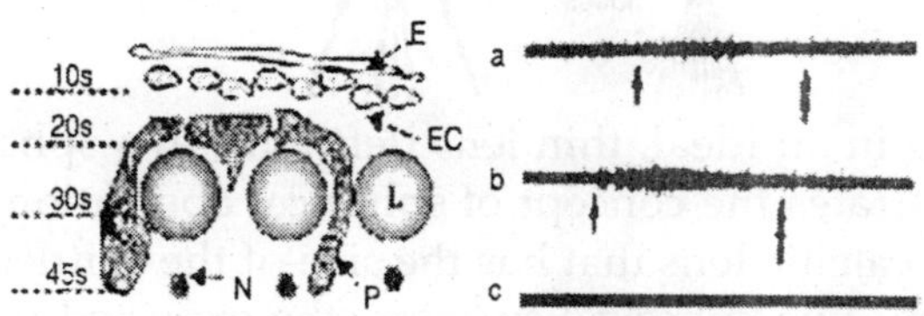

Left: schematic presentation of the structure of the epidermis and dermis in the DAP. Lenses (L), pigment (P), cuticle of the epidermis (E), epithelial cells (EC) and nerve bundles (N) are indicated.

Right: recordings of the nervous response. Single arrows represent "Light 'ON'". Double arrows represent "Light 'OFF'".

(a) Response to light in a control, untreated specimen or specimen treated for <10 s.

(b) Response to light when the arm was bleached for 30 s.

(c) Response to light when the arm was bleached for >45 s.

In technical terms, the brittlestar microlenses is a tunable optical device that exhibits a wide-range transmission tunability achieved by controlled transport of radiation-absorbing intracellular particles.

Other functions of the chromatocyte pigment may include the diaphragm action, and therefore, numerical aperture tunability; wavelength selectivity; minimization of the 'cross-talk' between the lenses, and therefore, the improved angular selectivity.

Aberrations

In an ideal lens, all rays of light would converge to the same point in the focal plane, forming a clear image. The influences that cause different rays to focus to different points in real optical systems are called aberrations.

For a most common lens type, i.e. for lenses made with spherical surfaces, rays that are parallel to the optical axis but at different distances from the lens centre do not converge to the same point. Rays that strike the outer edges of the lens are focused closer to the lens than rays that strike the inner portions of the lens. This effect, called spherical aberration, results in a considerable image blurring and presents a serious technological problem in lens fabrication.

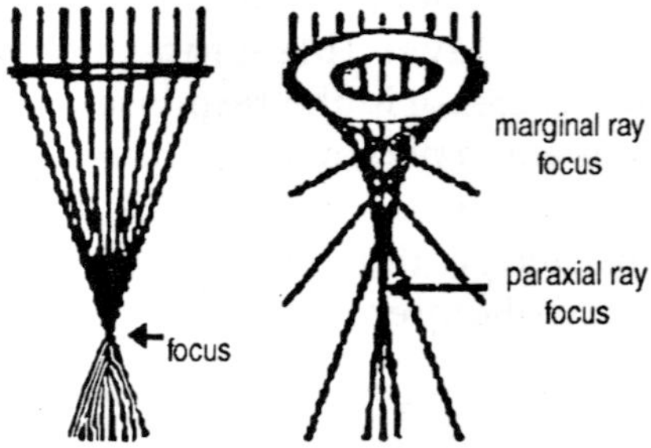

Ray tracing in an ideal, thin lens (left) and in a spherical, thick lens (right) that illustrates the concept of spherical aberration.

For a thick calcitic lens that has the size of the brittlestar microlenses and is formed by two spherical surfaces, the expected value of the light enhancement factor at the distance d from the lens would be E0 ~ 3-4. The experimental value of E determined in the lithographic experiments was 15 times higher than the latter. Therefore, the brittlestar microlenses must be significantly compensated for spherical aberration. Indeed, a close examination of the design of the brittlestar microlenses showed that they have a very peculiar shape.

Only the top surface is spherical. The bottom surface has a characteristic aspherical form. The calculated shape of the ideal calcitic lens that is totally compensated for spherical aberration appeared to completely coincide with the shape of the bio-lens. Moreover, the presence of pigment-filled chromatophores in the pores around the lenses would presumably block the light striking the outer portion of the lens, thus improving the operational size of the lens (L0).

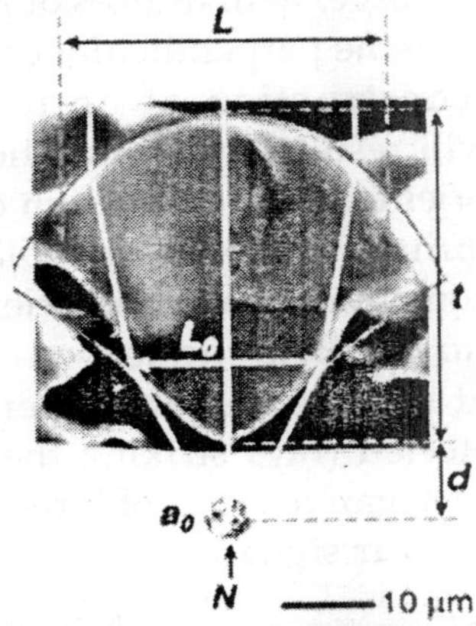

SEM of a cross-section of the brittlestar lens, showing its doublet structure delineated by one spherical and one aspherical surface. The nerve bundle (N) is shown under the lens. The bio-lens is superimposed with the calculated profile of an ideal lens that is compensated for spherical aberration (red contour). The match is striking.

Technological ways to minimize spherical aberration include:

(i) The use of lens doublets instead of the single lens.

(ii) Bending one lens of the doublet into its best, aspherical (usually parabolic) form.

(iii) The use of screens that disable the most problematic, exterior portions of the lens.

(iv) The use of high-index materials. Brittlestars seem to employ all of the above four approaches in their lens design. As a result, a remarkable level of compensation for spherical aberration is achieved and significantly enhanced light converges into one point where the photoreceptor is positioned.

Birefringence

Birefringence is the division of light into two components, Which is found in materials that have two different indices of refraction in different crystallographic directions.

It is observed in crystals that have a crystallographic axis of higher symmetry, so-called uniaxial crystals. The direction of the axis of higher symmetry is optically unique, in that the propagation of light in this direction is independent of its polarization.

This direction is called the optical axis of the material. Light propagating in any other direction will split into two beams that travel at different speeds.

Calcite is the classical example of a doubly refracting material. Its birefringence is extremely large, with indices of refraction of 1.66 and 1.48 along the optical axis and in the perpendicular direction, respectively. The use of calcite in the construction of optical lenses is, therefore, disadvantageous as it will result in the formation of two images unless the crystal is precisely oriented in the direction of the optical axis.

Amazingly, in the brittlestar dorsal arm plate it is indeed the case: the optical axis of the constituent calcite is oriented parallel to the lens axis and perpendicular to the plate surface.

Since only the light striking the array perpendicular to the plate surface is effectively detected (rays striking the surface at an angle are stopped by pigment), the negative effect of birefringence is corrected and the receptor receives one clear signal.

a

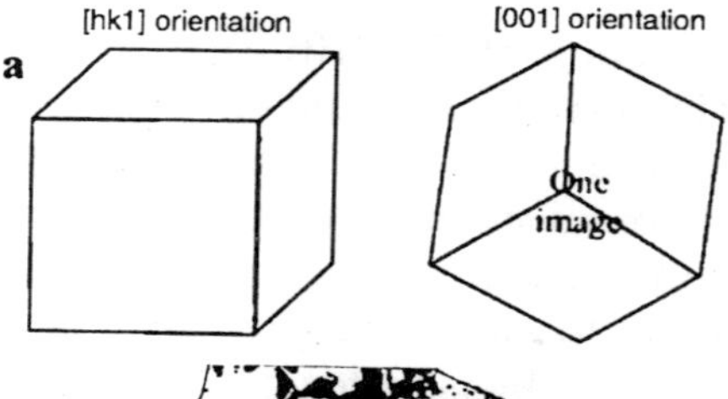

b

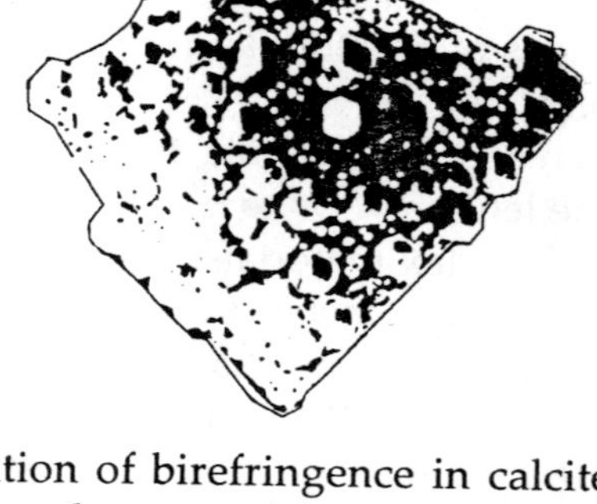

Schematic illustration of birefringence in calcite crystals. A double image is formed when the crystal is oriented in the general {h k l} crystallographic direction. One image is formed when the crystal is oriented in the [001] direction, i.e. along the optical axis. (b) SEM of the

DAP decorated with synthetic calcite crystals that grew epitaxially on the surface. The experiment visualizes the crystallographic orientation of the biogenic substrate. The overgrown calcite crystals and therefore the lenses are oriented in the optical axis direction.

Mechanical Properties

Calcite is typically used in echinoderms for structural purposes. This material is, however, highly brittle: cracks propagate easily along the {104} cleavage planes. Organisms have evolved several sophisticated means to reinforce this intrinsically brittle material for skeleton construction.

To that end, living organisms commonly employ organic- inorganic composite structures. Addadi and Weiner and co-workers suggested that specialized, intracrystalline macromolecules found inside biogenic calcite crystals of different origin are involved in the control of their mechanical properties.

These macromolecules are often adsorbed onto selective crystallographic planes in the crystal. It has been shown that in echinoderm calcite, the intracrystalline macromolecules interact specifically with crystallographic planes approximately parallel to the optical axis. Positioned oblique to the cleavage planes (the optical axis forms an angle of about 45u with the cleavage planes), these macromolecules provide an effective crack-stopping mechanism by absorbing and deflecting the advancing cracks.

Such a reverse fibre-reinforced composite exhibits reduced brittleness, increased plasticity, and fracture toughness.

As a result, the ossicles and associated microlenses formed by brittlestars are mechanically strong. It is also conceivable that intracrystalline macromolecules are involved in the regulation of the refractive index of calcite, thus further reducing the aberrations in the lens.

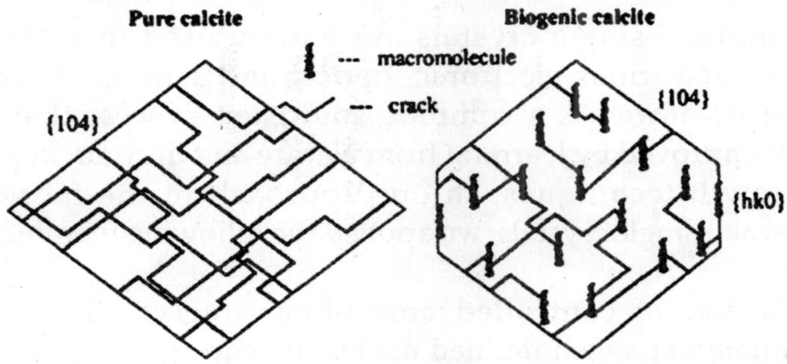

Schematic presentation of the crack-arresting function of specific intracrystalline macromolecules in biogenic calcites. See text for details.

In summary, photosensitive brittlestars are impressively armed for light sensing. They form an array of nearly perfect optical lenses that are micron scale, lightweight, mechanically strong, aberration-free, birefringence-free, and individually-addressed; they show a unique focusing effect, signal enhancement, intensity adjustment, angular selectivity, and photochromic activity.

Together with neural receptors and intraskeletal chromatophores, these microlenses form an effective optical device that may function as a compound eye.

Given appropriate neural integration, each DAP would have an effective visual field of ca. 10u, surveying a different part of the object space. Since O. wendtii has a large number of differently oriented DAPs, it could potentially extract a considerable amount of visual information about its environment.

Bio-inspired Engineering

For the purpose of advancing the state-of-the-art optical technology and striving towards the construction of a new generation of open space devices with a variable field of view, tunable transmission, wavelength selectivity, etc., the brittlestar microlens arrays represent an inspirational example. We believe that the above biological principles, if understood correctly and creatively applied in technology, could well revolutionize our ability to make tunable, lightweight, porous microlens arrays for a wide variety of applications.

The following fabrication strategies were inspired by the principles involved in the formation of echinoderm calcitic structures.

An interesting feature of the brittlestar microlens arrays is the presence of pores surrounding the lenses. These pores are important functional elements of the biological optical device, since they are used for the transport of pigment that regulates the illumination dose reaching the lens.

Microporous single crystals are widely used in technology as components of various electronic, optical and sensory devices. Their fabrication is, however, a complex, multistep process that could be potentially improved by learning from Nature and introducing biological crystal growth techniques. In our approach to the fabrication of micropatterned single crystals, we applied the following biomineralization principles:

(i) The use of controlled amorphous-to-crystalline transition initiated at a well-defined nucleation site.

(ii) The use of an ornate reaction volume that determines the shape of the deposited structure. This new bio-inspired crystal engineering strategy made it possible, for the first time, to directly fabricate millimeter-size perforated single crystals with a predetermined sub-10-micron pores and controlled crystallographic orientation.

We have also developed a novel, simple approach that uses multi-beam interference lithography to create porous hexagonal microlens (1-8 mm in diameter) arrays from photoresist materials. In our experiments, we used a continuous wave diode-pumped solid-state laser to photopolymerize a negative-tone resist.

The physical basis for the process is detailed below. When the interference light is transferred into a negative-tone photoresist during exposure, a periodic pattern of strongly and weakly exposed regions is generated.

The highly exposed regions are then polymerized and the unexposed regions are dissolved away to reveal the holes. When the difference between the adjacent strongly and weakly exposed regions is close to a certain threshold value, the gradual change in the intensity between the regions produces the lens-like topography in the photoresist film combined with holes.

The appearance of thus synthesized microlens arrays is strikingly similar to their biological prototype.

The lens size, shape, symmetry and connectivity are controlled by beam wave vectors and their polarizations, while the pore size is adjusted by laser intensity, exposure time and the concentration of a photosynthesizer in the resist. These synthetic microlenses are capable of focusing light.

The incorporation of holes in the lens array provides means for the transport of photoradiation-absorbing liquids and, therefore, for transmission and numerical aperture tunability and wavelength selectivity. These exciting preliminary results show that synthetic microlens arrays, inspired by the design of their nearly perfect biological prototypes, could be potentially used as highly tunable optical elements for a wide variety of applications.

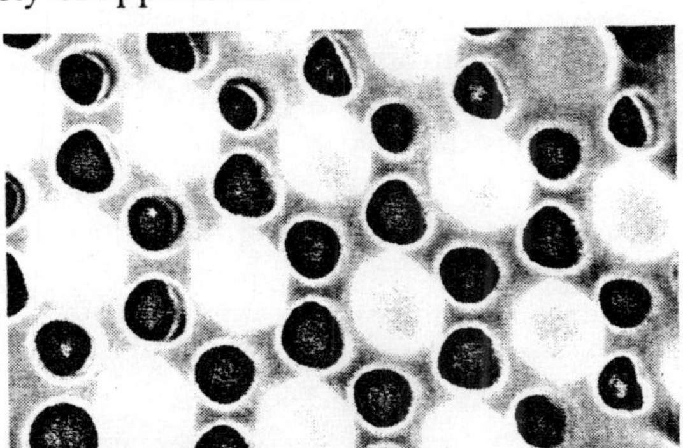

SEM of a synthetic biomimetic microlens array with integrated channelled pores fabricated by multi-beam lithography.

Multidisciplinary groups involving materials scientists, chemists, physicists, biologists work together trying to understand the mechanisms controlling the formation of elaborate structures of biological minerals. We believe that further studies of biological systems will increase our understanding of how organisms evolved their sophisticated optical structures for survival and adaptation and will provide additional materials concepts and design solutions.

Ultimately, these biological principles will improve our current capabilities to fabricate optical elements and contribute to the construction of novel, adaptive, micro-scale optical devices.

Chapter 16

Frictional, Voltaic or Galvanic, and Electro-magnetic Electricity, Wave Mechanics

THREE ELECTRICAL SOURCES

It has been found that there are three kinds of electricity, or, to be more accurate, there are three ways to generate it. These will now be described.

When man first began experimenting, he produced a current by frictional means, and collected the electricity in a bottle or jar. Electricity, so stored, could be drawn from the jar, by attaching thereto suitable connection. This could be effected only in one way, and that was by discharging the entire accumulation instantaneously. At that time they knew of no means whereby the current could be made to flow from the jar as from a battery or cell.

FRICTIONAL ELECTRICITY

With a view of explaining the principles involved, we show in Fig. 1 a machine for producing electricity by friction.

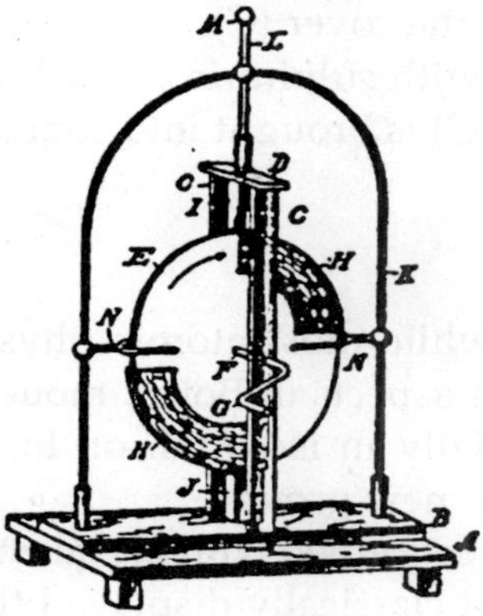

Fig. Friction-Electricity Machine

This is made up as follows: A represents the base, having thereon a flat member (B), on which is mounted a pair of parallel posts or standards (C, C), which are connected at the top by a cross piece (D). Between these two posts is a glass disc (E), mounted upon a shaft (F), which passes through the posts, this shaft having at one end a crank (G).

Two leather collecting surfaces (H, H), which are in contact with the glass disc (E), are held in position by arms (I, J), the arm (I) being supported by the cross piece (D), and the arm (J) held by the base piece (B). A rod (K), U-shaped in form, passes over the structure here thus described, its ends being secured to the base (B).

The arms (I, J) are both electrically connected with this rod, or conductor (K), joined to a main conductor (L), which has a terminating knob (M).

On each side and close to the terminal end of each leather collector (H) is a fork-shaped collector (N). These two collectors are also connected electrically with the conductor (K). When the disc is turned electricity is generated by the leather flaps and accumulated by the collectors (N), after which it is ready to be discharged at the knob (M).

In order to collect the electricity thus generated a vessel called a Leyden jar is used.

LEYDEN JAR

The jar (A) is of glass coated exteriorly at its lower end with tinfoil (B), which extends up a little more than halfway from the bottom. This jar has a wooden cover or top (C), provided centrally with a hole (D). The jar is designed to receive within it a tripod and standard (E) of lead. Within this lead standard is fitted a metal rod (F), which projects upwardly through the hole (D), its upper end having thereon a terminal knob (G). A sliding cork (H) on the rod (F) serves as a means to close the jar when not in use. When in use this cork is raised so the rod may not come into contact, electrically, with the cover (C).

The jar is half filled with sulphuric acid (I), after which, in order to charge the jar, the knob (G) is brought into contact with the knob (M) of the friction generator

WAVE MECHANICS

In the early 1920's while most atomic physicists were concerning themselves with different aspects of Bohr's model of the atom, Louis de Broglie, working essentially in isolation on his doctoral thesis at the Universite de Paris, broke new ground.

Louis de Broglie was concerned with the question of the wave-particle duality of light. Light had classically displayed the behaviour of a wave which the diffraction and interference phenomena studied by Fresnel and Young had revealed.

Einstein's description of the photo-electric effect, and the subsequent discovery of the Compton effect, however, had also revealed the particle nature of light. De Broglie tinkered with the notion that the photon might possess an unobservingly small mass.

This perhaps led him to speculate on the possible wave-particle dual nature of elementary particles such as electrons. De Broglie dropped the notion of a photon with mass. He proposed, however that particles like electrons might possess wave behaviour. He based his conjecture on the fact that the photon was both a wave and a particle. Why shouldn't the same be true for an electron?

To determine the frequency and wavelength of the electron wave, de Broglie borrowed directly from Planck and Einstein's concept of the photon. According to their quantum hupothesis, the energy of a photon is equal to Planck's constant, h, times its frequency, f.

$$E = hf$$

De Broglie assumed the identical relation held for the electron and hence, the frequency, f, of an electron is equal to its energy, E, divided by h. The momentum of the photon

$$p = E/c = hf/c$$

where we used the definition cT = c/f (T = 1/f) is the period of the wave. Although the relation between momentum and energy for the electron is not the same as the photon, dè Broglie assumed that the relation between momentum and wavelength are the same. He, therefore, concluded that the wavelength of the electron, ?, was also equal to h/p.

De Broglie applied his hypothesis of the wave nature of the electron to the problem of Bohr's atom and in particular to the question of how stable orbits could be formed. De Broglie assumed that the electrons orbiting the nucleus of the atoms formed standing matter waves.

Let us digress for a moment and consider the standing waves of a violin string which is pinned down at its two ends. The violin string can only vibrate in certain modes called standing waves. The condition on the vibration is that an integer number of half-wave lengths fit into the length of the string. Only these vibrations will reinforce themselves, after they are reflected from the ends of the strings. Vibrations with different wave lengths will interfere with their reflections from the end points and quickly die out.

De Broglie's application of his wave hypothesis to the electron enabled him to explain a key feature of Bohr's theory of the atom. His hypothesis regarding the wave nature of the electron applies to all particles. We must, therefore, ask why the wave nature of particles had not been observed before de Broglie made his hypothesis. Let us first consider the wavelength of a macroscopic object, for example, a baseball that has just been thrown by a player. The mass of the ball is approximately 200 g and its velocity is approximately 3 × 104 cm/sec. The wavelength of the ball, therefore is given by

$$\lambda_B = \frac{h}{mv} = \frac{6.6 \times 10}{200\ g 3 \times 10^4 cm/\sec} = 10^{-21} cm$$

With wavelengths as small as 10-21 cm one would never expect to detect any wave behaviour from a macroscopic object. It is only when we get to atomic size particles that we could possibly observe any wave behaviour because it is only for these small size objects that the wavelengths become large enough to detect wave behaviour. De Broglie, therefore, predicted that only subatomic particles such as electrons would display wave behaviour.

The next step in testing his hypothesis was an experimental observation of the wave behaviour of the electron. This did not happen immediately.

Although de broglie was able to explain Bohr's frequency condition, his contemporaries were quite skeptical about his result. They thought his work was a wild theoretical scheme unrelated to reality. There was even some doubt as to whether or not his work would be accepted for his doctoral thesis.

(His thesis was finally accepted and ironically turned out to be the only doctoral thesis which won its author both a doctor's degree and a Nobel prize). Luckily, de Broglie's work was brought to the attention of Einstein who was very favorably impressed by it.

Einstein, in turn, passed the thesis on to others. To Max Born he remarked, "Read it even though it might look crazy, it is absolutely solid." With Einstein's blessings the experimentalists began a systematic search to detect the wave nature of the electron suggested by de Broglie.

Several physicists around the world began working on the problem. One of the groups consisted of Davisson and Germer. They were attempting to detect the wave behaviour of particles by scattering electrons off of nickel. They were off to a flying start when their apparatus blew up and they had to start practically from scratch. Luckily they were able to salvage their target by treating it with heat. Their misfortune was actually a blessing in disguise. As a result of the heating, the nickel target had crystalized.

When Davisson and Germer returned to their experiment, they immediately observed a diffraction pattern identical to the one that had been observed in earlier X-ray scattering experiments. There was no doubt about it. The electron behaved exactly like a wave. It produced the same diffraction pattern as X-rays.

The paradox of quantum physics had come full circle; not only did waves display particle behaviour but now particles displayed wave behaviour.

The wave behaviour of the electron was displayed in other experiments following Davisson and Germer's success. The wave behaviour of neutrons was also discovered in a similar diffraction scattering experiment with a crystal.

Long before Davisson and Germer had completed their experimental demonstration of the wave nature of the electron, an Austrian physicist named Erwin Schrodinger began developing de Broglie's ideas of standing matter waves. De Broglie's work had been brought to his attention by Einstein whom he thanks in a letter dated Zurich, April 23, 1926: "Besides, the whole thing would certainly not have originated yet, and perhaps never would have, (I mean not from me), if I had not had the importance of de Broglie's ideas really brought home to me by your second paper on gas degeneracy."

Schrodinger developed de Broglie's notion of standing matter waves to a much greater extent. De Broglie's standing waves were essentially one dimensional circular waves.

Schrodinger considered the electron as a cloud which filled the entire space around the nucleus of the atom and vibrated as a three dimensional standing wave. Schrodinger developed a wave equation to describe the electron which took into account the electromagnetic force exerted by the proton on the electron.

There is an amusing story told by Dirac related to Schrodinger's discovery of his famous equation.

It seems that when Schrodinger was originally developing his ideas the very first equation he derived was a relativistic one. This equation has a number of complications. Consequently, when Schrodinger applied this equation to the problem of the hydrogen atom he was unable to obtain the desired experimental results.

He was greatly discouraged and depressed and dropped the whole project for a number of months. When he returned to his work he realized that if he dropped the relativistic equation and ignored relativistic effects due to the electron's motion that a non-relativistic reversion of his original equation produced the desired results.

Schrodinger's equation is a non-relativistic equation. Relativistic corrections to his equation were developed later by Dirac who discovered the source of Schrodinger's original problems. Schrodinger encountered difficulties because he had attempted to make two steps at one time. This rarely happens in physics. progress is usually made one step at a time. Schrodinger developed de Broglie's idea, who in turn, had developed Einstein's idea, who, in turn had developed Planck's idea.

De Broglie's standing wave formalism only allowed him to reproduce the Bohr frequency condition. Schrodinger's equation allowed him to calculate the exact energy of the atomic levels. Because his equation included the effects of the electromagnetic potential on the electron, he was also able to derive the corrections to the Bohr levels due to an external magnetic field (Zeeman effect) or due to an external electric field (Stark effect).

In addition to giving excellent agreement with spectral data Schrodinger's results helped to explain the nature of the electron's behaviour in the atom. instead of the picture of the electron jumping discontinuously from one quantum state to another, a new view developed.

An atomic transition from one level to another was seen as a transition from one standing wave configuration to another. The image of the standing wave also helped to explain why the orbits were quasi-stable and why only certain orbits were allowed.

As with any other new development in physics, Schrodinger's results provided the solution to a number of problems but only at the price of raising new problems. The foremost question was the interpretation of the standing waves.

What were they in fact waves of? Schrodinger at first considered the electron literally as a material wave whose dimensions were given by that of the standing wave. it became apparent, however, that the standing wave represented a cloud of probability. That the actual dimensions of the electron were quite small and that only its probability of being detected was spread out through space.

These probability standing waves are displayed for the hydrogen atom. These patterns are not direct photographs of the atom but only represent the patterns of the probability waves. Max Born was the first to make the probabilistic interpretation of Schrodinger's results.

He was most likely influenced by the 1924 work of Bohr, Kramer and Slater in which they claimed that the electromagnetic wave represented the probability of detecting a photon. Schrodinger resisted the probabilistic interpretation at first because he felt that he had eliminated the discontinuity of quantum jumps by considering the electron as a material wave. After Bohr had finally convinced him that his theory was correct but his interpretation not valid, Schrodinger remarked in frustration, "If one has to stick to this quantum jumping, then I regret ever having gotten involved in this thing".

Six months prior to Schrodinger's developments in the summer of 1925 Heisenberg independently developed a completely different approach to atomic theory. His mathematical description of the atom known as matrix mechanics also dealt with probabilities.

Heisenberg argued that atomic theory should only deal with observable i.e., quantities that can be directly measured. He, therefore, developed equations for the probabilities that an atom would make a transition form one quantum state to another.

Heisenberg formulated his equation in the quantum domain for which Bohr's correspondence principle was valid. This enabled him to exploit classical physics which is still valid in this domain. Using his equations, Heisenberg was able to calculate correctly the probabilities of transitions from one atomic level to another, as well as the energies of each level.

Schrodinger was able to show that Heisenberg's matrix mechanics were mathematically identical. Schrodinger's formulation of quantum mechanics proved to be more convenient for actual calculations. Heisenberg's contribution was just as important, however. Although calculations within the Heisenberg's matrix mechanics were clumsier, his scheme proved extremely useful from a theoretical point of view. The relativistic formulation of quantum mechanics to be discussed later were developed by Dirac using the Heisenberg picture. Heisenberg's matrix mechanics like Schrodinger's wave mechanics were non-relativistic.

Heisenberg's formulation of quantum mechanics also leads naturally to the Heisenberg uncertainty principle. This principle has played a crucial role in understanding the physical ideas behind quantum mechanics and has in itself led to a number of developments in atomic physics.

The uncertainty principle states that it is impossible to make an exact determination of both the momentum and position of a particle no matter how accurately they are measured. The uncertainty principle follows from Heisenberg's equations.

Once formulated, it is easy to demonstrate that it arises from general considerations of measuring atomic phenomena and indeed, explains the necessity of a probabilistic description of atomic processes. We shall return to this question later, but let us continue our description of the uncertainty principle.

The uncertainty principle states that there is an intrinsic built-in theoretical limitation to how precisely one may measure both the position and the momentum of a particle. It is possible to measure either the momentum or the position of the particle as accurately as one cares to, however, one's measurement of the other variable suffers as a consequence.

For instance, it one reduces the uncertainty in one's measurement of the particle's momentum one automatically increases the uncertainty in the measurement of its position. Labeling the uncertainty in the measurement of the momentum, p, and the position, x, by ? p and ? x respectively, the mathematical expression of the uncertainty principle takes the following form: the product of the ? p times ? x is always greater than or equal to h, Planck's constant, i.e.,

$$p\,x - h$$

In fact, if one knows the momentum precisely such that p = 0 then x becomes infinite which means one loses all information about the position.

Heisenberg also showed that the uncertainty principle also applies to the measurements of the energy, E, and the lifetime, t, of a system. The measurement of one interferes with one's knowledge of the other. If ? E and *t* are the respective uncertainties of the energy and time measurement then the uncertainty principle states that the product of these uncertainties will always be greater than or equal to h.

We shall, for the most part, consider the uncertainty principle as applied to position and momentum but there will be some occasion to refer to energy and time as well.

Many physicists and lay thinkers found the uncertainty principle a complete anathema. They could not conceive how theoretical limitation to measurements could possibly be imposed upon physics. They were also offended by the probabilistic nature of the quantum mechanics which the uncertainty principle seems to epitomize.

The uncertainty principle, to my way of thinking, on the other hand, represents a natural limitation to the study of microscopic quantities whose energy is quantized. The uncertainty principle helps one understand why a probabilistic description of atomic processes is necessary.

In order to describe a physics system which, after all, is the object of physics, it is first necessary to observe or know the system, i.e., to be able to gather information about the system. In order for the information to be useful, we would like to be certain that the act of making one measurement on the system does not alter it to the extent that we are no longer dealing with the same system.

Otherwise, we will collect information of successively different systems and never be able to describe the original system. For example, suppose we wish to know both the position and momentum of a body at the same time. If each measurement of the position imparts some unknown momentum to the body we will never be able to measure both its momentum and position simultaneously. For the measurements of macroscopic bodies this has never been a problem. One could always arrange to measure the position of a large body without affecting its momentum.

Let us consider the determination of the position and momentum of an automobile, for example. If we were to determine the position of an automobile by crashing another automobile into it then I would certainly change the original auto's momentum in some undetermined manner making the precise measurement of its original momentum impossible. But we do not have to make my measurement in such a heavy-handed manner.

For instance, we could throw a tennis ball at the car and the change in the momentum we would produce would almost completely negligible.If we wish to be even more discreet about my measurement we can make my measurement of the car's position visually. Even in this case, we will affect the car momentum ever so slightly since a visual measurement involves bouncing light off the car into my eyes.

As we know, light carries momentum so even in this case we impart some unknown momentum to the car. This effect is completely negligible when one takes into account that the momentum of the car is 1030 times the momentum of a photon of visible light.

This can also be seen by examining the uncertainty principle mathematically for the case of an automobile whose length, mass and velocity are approximately 300 centimetres, one million grams and 3000 cm/sec respectively.

If we divide the uncertainty evenly between the momentum and the position, the uncertainty principle prevents us from measuring the length of the car or its velocity more accurately than one part in 1019. Since the accuracy for making measurements is much less than this and since the accuracy needed to describe a system does not have to be anywhere near one part in 1019 the uncertainty principle has absolutely no effect of the description of a macroscopic system like an automobile.

Let us now consider the limitations the uncertainty principle imposes for the case of an electron in an atom. Planck's constant will no longer be such a small number.

Instead of distances, like 300 cm for the automobile, we must now consider distances of the order of 10-8 cm. Instead of a mass of 106 g the mass of the electron is .9 × 10-27 grams, and hence, its momentum is considerably less than that of the automobile. For an electron with a velocity of.1 c the product of its position times its momentum (10^{-8} cm × .9 × 10^{-27} gm × .1 × 3 × 10^{10} cm/sec) is approximately 30 × 10^{-27} gcm^2/sec or just 5h.

It is clear that the uncertainty principle imposes severe limitations on how accurately and the momentum and position of an electron may be determined since the uncertainty is the same order of magnitude as the quantities to be measured.

Let us consider physically what is involved in determining the position and momentum of an electron. In actuality, it is not much different than measuring the position and velocity of an automobile by crashing another automobile into it.

There are no particles smaller than an electron. Therefore, if we wish to detect the electron using another particle the best we can do is to use another electron. We can not chop an electron into a thousand pieces and use a tiny chunk of an electron as a detector. We are obliged to use another electron or a larger particle.

The only other alternative is to use a photon. This presents problems as well because the photon also carries momentum. This problem can be minimized by choosing to use a low energy and consequently a low momentum photon. The only difficulty with a low momentum photon is the fact that it will have a large wavelength,

$$\lambda = h/p$$

$$\Delta p.\,\Delta x = \lambda.\frac{h}{\lambda} = h$$

Thus, we see in accordance with the uncertainty principle that is the minimum value for the product of the uncertainties of the position and the momentum. There is no way of avoiding the uncertainty principle.

In discussions with students it is sometimes suggested that the uncertainty in momentum introduced by the momentum of the photon can be avoided by shooting photons at the particle equally from all sides. This ingenious proposal would avoid the uncertainty in momentum. However, in order to direct the photons equally on all sides one would have to know the position of the particle. If we could do this we wouldn't have to make the measurement in the first place.

The uncertainty principle is not really such a mysterious concept. In simple terms, it states that in order to obtain information about a system, it is necessary to disturb the system making it impossible to ever obtain a complete set of information about the system. To know something is to interact with it and hence, change it.

If we apply this concept to other areas of human study it will seem even less mysterious. Let us consider the problem of a biologist studying the behaviour of a group of animals. The biologist is well aware that his observations will be contaminated to some extent since the animals will be aware of being observed and behave differently than they would in a total state of nature.

One can minimize the effects of such observations. Jan Goodall's observations of chimpanzees made while she lived in the wilds with her subjects is far more accurate than those made by observing chimpanzees in a zoo. In spite of all her care and trouble, the behaviour of the chimpanzees observed by Jane Goodall was still influenced to some extent by her presence. The concept of the uncertainty principle can also be applied to interpersonal relations. if I want to get to know a person, then I must interact with them and hence, change them to some extent.

The limitations of the physicists' knowledge of his physical world arises more or less for the same reason that they arise in the field of biology of interpersonal relations. To know something is to interact with it and hence, disturb it. Perhaps the same reason that so many people (both scientist and lay people) find the uncertainty principle so disturbing is that a myth has developed surrounding the physicist.

As a result of the success of Newtonian physics people began to believe that the physicists were able to provide an exact mathematical description of the physical world. The success of the post-Newtonian physics in describing other physical phenomenon such as electricity, magnetism, heat, sound and light help to reinforce this myth.

Thus, it was a great shock to scientists and non-scientists alike when the study of atomic physics revealed that there are limitations to man's knowledge. We shall return to some of the philosophical implications of the uncertainty principle raised here but let us first examine its implications for physics.

In classical or Newtonian physics it is possible to determine the exact position and momentum of a body. Once this information is known it is then possible to specify the exact position and momentum of the body for all future times as long as one knows the forces acting on the body. This programme is no longer possible within the framework of quantum mechanics even is one knows all the forces on the body. The uncertainty principle only permits a partial or approximate knowledge of the location and momentum of the body.

The uncertainty principle forces a change in our description. Instead of specifying the exact location and momentum of the particle as a function of the time as was done in classical physics, we are now forced to describe the particle in terms of the probability of finding it at some place, x, with a momentum, p. Within the framework of quantum mechanics, the probability of determining the particle's position and momentum is described quantum mechanically by the probability amplitude or wave function, which has a unique value for each position in space x, y and z.

The equations describing the behaviour of the wave function, is the Schrodinger equation referred to earlier. When Schrodinger first discovered his equations he did not connect the wave function, (x, y, z) with probability but rather interpreted phi as the density of the electron cloud which he considered to be spread out through space. It was max Born who pointed out that the correct interpretation was to continue to assume that the electron is a point particle and to regard as a measure of the probability of detecting the point electron at some point in space. He showed, in fact, that the probability of finding the electron located at the point in space x, y and z is just the absolute value of the wave function multiplied by itself, $(x, y, z)^2$.

He also showed that Schrodinger's equation determines the probability that the particle has a particular value of the momentum can also be determined from but involves a more complicated mathematical operation than multiplying by itself. These mathematical details need not concern us. The important point is that once one knows a particle's wave function, (x, y, z), one can determine the probability it will have a particular momentum and position.

The wave behaviour of the electron is due to the fact that its probability amplitude, behaves like a wave. The Schrodinger equation is basically a wave equation. The diffraction pattern observed when a beam of electrons passes through two slits is due to the interference of the wave functions passing through the two slits.

It is not due to the interference of a beam of electrons passing through one slit with the beam of electrons passing through the other slit. This has been experimentally demonstrated by reducing the flux of electrons so that only one electron at a time passed through the two slit system.

The position on the screen where the electron landed after passing through the slits is recorded and tabulated as illustrated. After a sufficient time passes the pattern that emerged was the same characteristic diffraction pattern one obtained with a high flux beam.

There were positions on the screen where electrons would go only if both slits were open and would not go it slit 1 was open and slit 2 was closed or if slit 1 was closed and slit 2 was open. There were also positions where the electrons would not go if both slits were open but would go if one or the other slit was open.

This is extremely mysterious. In the latter case I open slit 1 and close slit 2 and observe electrons at position Y on the screen. If I now open both slit 1 and slit 2, electrons no longer go to position Y.

It is impossible to understand in terms of a particle how the opening of slit 2 suddenly prevents electrons from going to position Y via slit 1. In view of the fact the results hold even with one electron at a time the only way of interpreting this is to assume that the electron passes through both slits and interferes with itself.

But how is this possible if the dimensions of the electron are smaller than the distance between the two slits? The way this phenomena is understood in terms of our quantum mechanical description is to recognize that the probability amplitude, describing the electron is interfering with itself.

WAVE PACKETS

De Broglie was the first to indicate the wave-like properties of the electron which explains the diffraction behaviour described above. De Broglie assigned to the electron a single wave length ? related to its momentum p by

$$\lambda = h/p$$

We have learned, however, in our discussion of the uncertainty principle that a particle rarely has a precisely defined momentum. We therefore expect the wave function describing the particle to be composed of many waves, each with a different wavelength. Such a superposition of waves is called a wave packet. The momentum of these waves and hence the particle is spread out between h/? and h/? + ? ?. The uncertainty in the momentum of the particle is therefore approximately

$$\Delta p. = \frac{h}{\lambda} - \frac{h}{\lambda + \Delta\lambda} \approx \frac{h}{\lambda^2}\Delta\lambda$$

A well-known property of waves is their ability to interfere with each other both destructively and constructively. This happens with the waves in the wave packet. The probability of finding the particle is highest where the waves interfere constructively and is negligible where they interfere destructively.

The size of the wave packet in space is therefore determined by where the waves interfere Since the probability of finding the particle is greatest between X and X + ΔX where the waves are not interfering destructively, the position of the particle is approximately X with an uncertainty just equal to X.

We note that the products of the uncertainties in momentum and position p X is just equal to h, the minimum value allowed by the uncertainty principle.

The size of the wave packet describing an atomic particle increases with time. This does not mean that the actual size of the particle increases.

It remains the same but the wave packet and hence, the uncertainty in the particles position does increase with time. The reason for this is that the different waves in the packet are travelling at different speeds because of their spread in momentum.

The waves with momentum p + p gets ahead of the waves with momentum p and hence, there is an increase in the size of the wave packet. This surprising result is a unique feature of quantum mechanics not to be found in classical physics

BARRIER PENETRATION

Perhaps the most baffling quantum mechanical effect of all is barrier penetration or tunnelling. Let us consider a particle under the influence of a spherically symmetric force. The potential energy of the particle under the influence of the force. The nature of the force is to create a barrier which prevents particles with total energy less than Vo from entering or leaving the region r less than r0.

From a classical point of view if a particle's total energy is less than V0, the maximum potential energy of the force, it will remain either exclusively in the region $r < r_0$. It will not be able to pass from one region to another for, in so doing, it would have to have a potential energy greater than its total energy, which is impossible.

The nuclear force which traps the protons and neutrons (nucleons) with in a nucleus is a force similar to the one which produces the potential energy distribution shown in Fig. 18. If the classical laws of physics applied, one would never expect a proton or neutron to escape the nuclear barrier. There are cases, however, of unstable nuclei where the nucleons escape the nucleus.

This effect can be explained in terms of quantum mechanics. Because ? is not identically zero in the forbidden region there is a very small but finite probability that the particle will enter this region. If the particle passes through this region to values of r r0 it will then be free of the nucleus because now its total energy is greater than the potential energy and hence, its wave function in this region no longer needs to remain small. In this way nucleons are able to tunnel through the barrier created by the nuclear force and escape the nucleus. An equivalent event on the macroscopic level would be the exit of a student from a school room by walking through the walls and leaving them intact.

The probability of the tunneling of a particle is very small because the wave function in the forbidden zone is so small. Given the potential energy due to a given force one can use the Schrodinger equation to calculate the probability of the penetration of this potential barrier. The calculation of the lifetime of unstable nuclei based on barrier penetration has given excellent agreement with experimental value confirming the quantum mechanical effect of tunnelling.

This effect is strictly quantum mechanical and would not be possible were it not for the uncertainty principle. When the particle passes through the forbidden zone, a violation of energy conservation takes place. The time during which this violation occurs is very small so that the product of the energy conservation discrepancy times the time the particle is in this state is less than h. The energy conservation violation can never be detected because measurements accurate enough to show this are not possible.

The diffraction of electrons, the expansion of wave packets and the tunnelling of potential barriers all involve breakdown of causality. Causality is the concept which forms the very foundation of classical physics. Even relativity which changed the notions of causality. In fact, causality plays an important role in the formulation of relativity. The breakdown of the concept of causality, intrinsic to the formulation of quantum mechanics by Heisenberg and Schrodinger, produced shock waves among physicists and philosophers which persist to this day.

Although the formulators of quantum mechanics were forced to give up the notion of casualtiy, they were still able to construct a theory of which possessed productive powers. They could no longer describe the behaviour of an individual particle. They were able to provide a probablistic description of individual particles which enabled them to predict the behaviour of statistically large ensembles of particles. Thus, they were able to preserve the most basic aspect of a scientific theory, namely its ability to make predictions which can, in turn, be verified by experiment. In fact, quantum theory was able to explain a great deal of the behaviour of atoms and molecules.

Perhaps the most important of these results was the description of the spectral and chemical properties of various atoms and molecules as well as the structure of matter such as solids and liquids.

The success of quantum physics as a descriptive theory of large numbers of atoms was universally accepted within the physics community. A schism developed within the community, however, regarding the interpretation and meaning of the theory. A number of scientists including Einstein, Schrodinger, Planck and Later, de Broglie, were unable to accept the operation of chance within the framework of quantum mechanics. They believed that since quantum theory could not provide a causal description of the behaviour of individual particles that it was an incomplete theory.

They conceded that quantum mechanics was a logically consistent scheme which was able to describe experimental results accurately. They accepted that the limitations of measurements imposed by the uncertainty principle made a probablistic description necessary. They believed, however, that the uncertainty principle was only a limitation of our knowledge and that, in actuality, the particle has both a precise position and velocity and that its behaviour is causally determined. They could not accept the notion that chance could actually enter into the behaviour of the physical world.

Einstein, who became the leader of this position, expressed the concern of his school of thought with his often quoted remark, "I do not believe in a God that plays dice." Einstein and the others believed in the existence of an underlying determinism which actually guided the particles. They simply did not accept the idea that the behaviour of particles could be governed by chance. They wanted to know what was really happening to the particles. They believed a theory would eventually emerge in which the hidden causality would appear. They, therefore, considered quantum theory as incomplete and anticipated the appearance of a fuller theory which would eventually replace it.

In opposition to this point of view were the proponents of the Copenhagen interpretation who had formulated their ideas in Copenhagen in the late 1920's under the leadership of Niels Bohr and included Heisenber, Dirac, Born, Pauli and Kramers. They regarded the new quantum mechanics as a complete theory. They believed that the uncertainty principle imposed a limitation on our knowledge of the world and hence, a limitation on the behaviour of the particles themselves. They made no distinction between reality and our knowledge of reality.

They considered the question asked by Einstein and his followers, 'what is really happening to the particle is meaningless?' What can not be observed or measured is on no concern to physics since there can never be verification of a theory of unobservable behaviour. An expression of the view is found in the writings of Dirac who wrote the following: "The only object of theoretical physics is to calculate results that can be compared with experiment, and it is quite unnecessary that any satisfying description of the whole course of phenomena should be given."

One should not interpret Dirac's statement in a trivial way and conclude that he and his colleagues decided that since their results explain experimental results there was no need for further reflection. The members of the Copenhagen school were just as perplexed and disturbed by the new ideas. They, too, spent long hours trying hard to understand the underlying ideas behind their new theory. The following passage from Heisenberg's book entitled "Physics and Philosophy" indicates the nature of their frustration: "During the months following these discussions an intensive study of all questions concerning the interpretation of quantum theory in Copenhagen finally led to a complete and, as many physicists believed, satisfactory clarification of the situation. But it is not a solution which one could easily accept.

I remember discussions with Bohr which went through many hours till bery late at night and ended almost in despair; and when, at the end of the discussion, I went alone for a walk in the neighbouring park, I repeated to myself again and again the question: Can nature possibly be as absurd as it seemed to us in those atomic experiments?"

Coming to grips with the peculiarities of quantum mechanics was no easy task for Bohr and his co-workers. Einstein and his sympathizers in a sense avoided the problem by refusing to give the new quantum mechanics the status of a complete theory. They put the problem off by claiming that some day a complete theory would arise and explain all. Bohr, on the other hand, accepted the paradoxes of quantum mechanics even though they contradicted the intuition of physics that he and others had developed from their study of classical physics.

Bohr recognized that although quantum mechanics had completely invalidated classical physics that it was still necessary to use the language of classical mechanics to describe the behaviour of atomic systems. This is perhaps the ultimate contradiction of quantum mechanics. It would be quite impossible to develop quantum mechanics in purely quantum mechanical terms. Thought by itself is a classical process. Our entire experience is classical. If the laws of quantum physics operated on the macroscopic level we would never have developed physics. Life itself would not be possible. imagine the difficult of crossing the street safely if the approaching cars are only 90% on the rad and 10% of the time are on the sidewalk.

The world of our experience if classical, not quantum mechanical. It is fairly obvious the only language that we have at our disposal to describe the world whether it is the macro-world or the atomic world is the language of classical physics, even if that language is inadequate. It should not surprise us, therefore, that our description of the atomic world will be less than satisfactory at times. The description of light and elementary particles such as electrons is a perfect example. In classical physics, the concept of a wave and the concept of a particle were well-defined notions which were mutually exclusive.

A wave represented a phenomenon whose character was completely different than that of a particle. Within the framework of classical physics these two concepts were used to describe light and electrons, respectively. Light was a wave phenomenon and the electron, a particle and that was that. Within the framework of quantum physics, the behaviour of light and the electron is a great deal more complicated than that of a wave and a particle, respectively. There is, in fact, no way of describing their behaviour classically. The best one can do is to say that there are times when light behaves like a wave and there are times when it behaves like a particle. The same is true of the electron. The classical notions of the wave and the particle, according to Bohr, form complementary descriptions of the electron. They also form a complementary description of light. Although the wave and particle notions mutually exclude each other they are both necessary for a description of either light or the electron. This is the essence of Bohr's complementary principle.

The classical notions of position and momentum are also complementary concepts. In classical physics, one can assign an exact value to both quantities whereas in quantum physics the more one learns about one variable the less one knows of the other. A quantum system can not have a well-defined position and momentum at the same time. Yet, these two mutually exclusive quantities are needed to describe the quantum system.

Position and momentum are complementary concepts in the same sense as the wave and the particle notion are complementary. There is a famous quote Bohr made regarding truth which I believe contains the essence of the complementary principle. He said: "[There are] two kinds of truths. To the one kind belong statements so simple and clear that the opposite assertions could not be defended. The other kind, the so-called "deep truths" are statements in which the opposites also contain deep truths."

Einstein and his followers could not accept the tenets of the complementary principle with its dichotomous description in terms of waves and particles. Instead, they adopted another dichotomy. They believed in two levels of reality, the one which is observed and the other which "really is". This division is not unlike the one Kant makes between phenomena or perceived events and noumena, or things in themselves. Einstein's position with regard to the completeness of quantum theory is that while quantum mechanics presently describes phenomena accurately enough, it is incomplete because it can not describe noumena. Only a causal description could properly describe the noumena. The Copenhagen school denies the distinction between phenomena and noumena and like the British empiricist, believes one can only describe that which is perceived.

Einstein believes that there is an intrinsic order within the physical world which exists outside of man. Bohr, on the other hand, believes that the order we have discovered in the physical world we have inposed through our mental activity. Einstein believes that the order was there and we were clever enough to uncover it. The difference in these two men's attitude partially explains their reaction to quantum theory. Einstein is disappointed because the order that quantum mechanics reveals is only partial, it is incomplete and hence, so is the theory. Bohr, however, seems quite delighted that he and his co-workers were able to go as far as they did and reveal as much order as they were able to. For Einstein, the proverbial half glass of wine is half empty whereas for Bohr it is half full.

Einstein and Bohr debated their respective positions with each other for the rest of their lives. Even after Einstein had died Bohr, in his mind, was still trying to answer the question raised by the man who refused to believe that God plays dice. It way always a source of personal frustration to Bohr that he could not win Einstein to his side.

The friendly debate that arose between the two men was like none other before in the history of science. In previous debates, the conflict usually centered on which of two conflicting theories provided the best description of nature. The goals and aims of science were usually agreed upon. In this debate, the reverse was true. Both sides agreed that quantum mechanics provides an accurate description of the experimental data. What was at issue was the aims of physics. The agreement as to whether or not quantum mechanics is a complete theory arose from a difference in opinion on the basic goals of science.

Einstein argued that the aim of science was to provide a causal description of nature while Bohr maintained that the only aim was to describe the experimental data. There was no way they could really resolve their conflict. All the same, they argued back and forth for many years. Einstein was continually proposing thought experiments to demonstrate possible contradiction in the theory. Bohr was always able to answer him satisfactorily Einstein persisted, but finally had to admit that the theory was self-consistent but would never concede that it was complete. Einstein's friend, the physicist Ehrenfest, one chastised him for his stubbornness "Einstein, shame on you. You are beginning to sound like the critics of your own theories of relativity.

Again and again your arguments have been refuted but instead of applying your own rule that physics must be built on measurable relationships and not on preconceived notions, you continue to invent arguments based on thos same preconceptions."

It is perhaps one of the great ironies of all physics that Einstein, the inventor of relativity, the greatest innovator physics had ever known was unable to accept quantum mechanics in spite of all his important contributions in the field. Bohr, probably better than any one else, understood the wisdom of the following remark attributed to von Helmholz, "The origination of a new concept finds, as a rule, that it is much more difficult to find out why other people do not understand him than it was to discover the new truths."

There is no true resolution to the Bohr-Einstein debate. The only way that the Einstein position could truly be vindicated is for someone to discover a complete theory which both describes the experimental data and restores causality. Numerous attempts in this direction have been made without any notable success. In the absence of such a theory, one is more or less inclined to agree with Bohr. Only time will tell.

Index

S

T

V

W

X